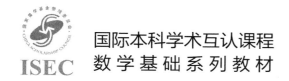

国际本科学术互认课程
数学基础系列教材

微积分 II（双语版）

程晓亮　华志强　王洋　陈丽莹　刘鹏飞　包丽平　敖恩　◎编著

图书在版编目(CIP)数据

微积分：双语版.Ⅱ/程晓亮等编著. —北京：北京大学出版社，2018.4
（国际本科学术互认课程·数学基础系列教材）
ISBN 978-7-301-29418-5

Ⅰ.①微… Ⅱ.①程… Ⅲ.①微积分—高等学校—教材 Ⅳ.①O172

中国版本图书馆 CIP 数据核字（2018）第 056924 号

书　　　名	微积分 Ⅱ(双语版)
	WEIJIFEN Ⅱ
著作责任者	程晓亮　华志强　王　洋　陈丽莹　刘鹏飞　包丽平　敖　恩　编著
责任编辑	曾琬婷　尹照原
标准书号	ISBN 978-7-301-29418-5
出版发行	北京大学出版社
地　　址	北京市海淀区成府路 205 号　100871
网　　址	http://www.pup.cn　新浪微博: @北京大学出版社
电子信箱	zpup@pup.cn
电　　话	邮购部 62752015　发行部 62750672　编辑部 62767347
印刷者	北京市科星印刷有限责任公司
经销者	新华书店
	889 毫米 × 1194 毫米　16 开本　11.25 印张　330 千字
	2018 年 4 月第 1 版　2024 年 1 月第 5 次印刷
定　　价	58.00 元

未经许可，不得以任何方式复制或抄袭本书之部分或全部内容。
版权所有，侵权必究
举报电话: 010-62752024　电子信箱: fd@pup.pku.edu.cn
图书如有印装质量问题，请与出版部联系，电话: 010-62756370

内 容 简 介

本书是根据"国际本科学术互认课程"(ISEC)项目对高等数学系列课程的要求,同时结合 ISEC 项目培养模式进行编写的"微积分"双语教材.全书共分 4 章,内容包括:空间解析几何初步、多元函数的微分、二重积分、无穷级数等.在内容选择上,既考虑到 ISEC 学生未来学习和发展的需要,又兼顾学生数学学习的实际情况,以适用、够用为原则,切合学生实际,在体系完整的基础上对通常的"微积分"课程内容进行适当的调整,注重明晰数学思想与方法,强调数学知识的应用;在内容阐述上,尽量以案例模式引入,由浅入深,由易到难,循序渐进地加以展开,并且尽量使重点突出,难点分散,便于学生对知识的理解和掌握;在内容呈现上,以英文和中文两种文字进行编写,分左、右栏对应呈现,方便学生学习与理解.

本书既可作为 ISEC 项目培养模式下"微积分"课程的教材,也可作为普通高等院校"微积分"课程的教学参考书,特别是以英文和中文两种语言学习和理解"微积分"的参考资料.

为方便教学,作者为任课教师提供相关的电子资源,具体事宜可通过电子邮件与作者联系,邮箱地址:chengxiaoliang92@163.com.

序　言

时值"国际本科学术互认课程·数学基础系列教材"第一部面世之际，本人在此向程晓亮老师和参加这套教材编写的各位 ISEC 教师表示热烈祝贺.

"国际本科学术互认课程"(International Scholarly Exchange Curriculum，简称 ISEC)项目，是国家留学基金管理委员会主持的、面向国内地方本科院校的教学改革项目.该项目致力于建设集国际化课程、国际化师资、国际教育资源于一体的国际化教育教学工作平台，并依托该平台，将具有国际先进水平的教学理念、教学思想和教学方法融入教师的教学实践，推动地方高校的教学改革.

本人深入参与了 ISEC 项目的两个基本环节：一是教师的课堂教学设计；二是教师和学生的明辨性思维训练.近距离观察了 ISEC 课程的教师和学生，有如下基本印象：教师培训成效显著.ISEC 教师对现代教育教学的思想和方法有相当程度的理解，对教师培训的内容有相当好的回应和反馈，对明辨性思维有相当程度的认知；ISEC 学生显示出不错的灵气和悟性.

ISEC 项目已经有坚实的基础，还有很大的发展空间，会有光明的发展前景.

EMI(English Media Instruction)是 ISEC 课程教学的特点之一，目前适合 EMI 教学的高等数学系列课程教材处于空白.为了适应 ISEC 课程教学的需要，解决 ISEC 学生高等数学的学习困难，在国家留学基金管理委员会 ISEC 项目办公室的关心和支持下，由程晓亮老师牵头，ISEC 项目院校吉林师范大学、内蒙古民族大学、贵州财经大学、贵阳学院、包头师范学院和赤峰学院等院校的 ISEC 教师参与，组织编写了一套英中对照教材——"国际本科学术互认课程·数学基础系列教材"，包括《微积分Ⅰ(双语版)》《微积分Ⅰ习题解析(双语版)》《微积分Ⅱ(双语版)》《微积分Ⅱ习题解析(双语版)》《线性代数(双语版)》《线性代数习题解析(双语版)》《概率论与数理统计(双语版)》和《概率论与数理统计习题解析(双语版)》.

这些 ISEC 教师因参与 ISEC 项目，了解了现代教育教学的思想和理念，掌握了课堂教学设计的方法和技巧，养成了明辨性思维的意识和习惯，并且将在 ISEC 项目中学习到的理念、思想和方法与自己的教学实践相结合，编写了这套双语教材，为推进 EMI 教学提供了有力的支撑.

已编写完成的《微积分Ⅰ(双语版)》，其语言显示出如下特点：教材中的英语不仅语言流畅，用词准确，而且充分兼顾了西方读者的思维习惯；教材中的中文与英文，不仅在空间形式上而且在内涵上形成了准确对应.更令人印象深刻的是，中文表述完全符合中国读者的思维习惯.可以说，这部双语教材有效地平衡了东、西方读者在思维习惯上的差异.

ISEC 项目高度重视教师和学生的明辨性思维素质的养成.这一意图在《微积分Ⅰ(双语版)》中得到不折不扣的贯彻：从基本概念的抽象到基本定理的证明，从基本思想的发展到理论体系的构建，无不体现出明辨性思维的特质.可以说，作者将明辨性思维有效地融入了教材的每一个章节甚至字里行间.因此，该教材是一部优秀的双语教材，也是养成学生明辨性思维素质的好教材.

另外,《微积分Ⅰ(双语版)》充分顾及学生认知发展的基本规律.德国伟大的数学家希尔伯特时常告诫自己的学生"从鲜活的案例开始",肯定了"鲜活的案例"在抽象数学概念、生成数学思想方面的重大作用.这部教材自觉遵循了希尔伯特的忠告,其案例的选择、内容的展开、理论的陈述,都遵循了由浅入深、由简单到复杂、由具体到抽象、由特殊到一般的基本原则.

随着ISEC项目的推进,我们期待能有更多这样的好教材面世.

殷雅俊

ISEC项目专家

清华大学教授

2017年6月于北京

前　　言

　　党的二十大报告对实施科教兴国战略、强化现代化建设人才支撑作出重大部署,明确指出:"教育、科技、人才是全面建设社会主义现代化国家的基础性、战略性支撑".青年强,则国家强.广大教师深受鼓舞,更要勇担"为党育人,为国育才,全面提高人才自主培养质量"的重任,迎来一个大有可为的新时代.在多种形式的人才培养途径中,要提升学生的国际视野,同时坚守中华文化立场,深化文明交流互鉴.这也正是编写出版"国际本科学术互认课程·数学基础系列教材"这套中英双语对照教材的初衷.

　　本书是"国际本科学术互认课程·数学基础系列教材"之一,它紧密结合国际本科学术互认课程(International Scholarly Exchange Curriculum,简称ISEC)对教学的要求,强调学习知识与训练思维的统一,强调教学理念与方法的统一,强调学习过程与学生能力提升的统一.在内容选择上,既考虑到ISEC学生未来学习和发展的需要,又兼顾学生数学学习的实际情况,以适用、够用为原则,切合学生实际,在体系完整的基础上对通常的"微积分"课程内容进行适当的调整,注重明晰数学思想与方法,强调数学知识的应用;在内容阐述上,尽量以案例模式引入,由浅入深,由易到难,循序渐进地加以展开,并且尽量使重点突出,难点分散,便于学生对知识的理解和掌握;在内容呈现上,以英文和中文两种文字进行编写,分左、右栏对应呈现,方便学生学习与理解.

　　本书作者都是经过多次ISEC教师岗前培训和专题培训的教师,并多次承担ISEC课程"微积分"的教学工作.正是在教学过程中,我们发现国内适应国际化教育教学需要的"微积分"双语教材匮乏.我们曾试图直接采用英文原版的"微积分"教材.但是,由于学生英文水平的限制以及以前没有双语学习的基础,特别是对"微积分"中涉及思想和方法的内容,学生把握起来比较困难.所以说,直接采用英文原版教材在某种程度上是不合适的.就"微积分"而言,国内有很多优秀的教材.然而,根据ISEC课程对学生发展的目标要求,是不宜采用中文教材的.正是在这样的背景下,我们结合多轮"微积分"教学经验,精心选材与设计,撰写了这部"微积分"双语教材.

　　全书由程晓亮、华志强、王洋撰写,参与编写、审阅、修改工作的还有陈丽莹、刘鹏飞、包丽平、敖恩.

　　在本书的编写过程中,我们得到了国家留学基金管理委员会ISEC项目办公室的大力支持.可以说,没有ISEC项目办公室的鼓励与支持,就没有这套教材的孕育,更谈不上这套教材的面世.吉林师范大学教务处各位领导十分关心这套教材的撰写,尤其是李雪飞教授给予了无微不至的关心与大力的支持.在此,我们表示衷心的感谢.

　　ISEC项目专家、清华大学教授殷雅俊在百忙之中为这套教材作序.借助ISEC教师岗前培训和专题培训,我们多次得到殷雅俊教授的培训与指导,其内涵丰富、思想深邃,使我们受益匪浅.在此,特别对殷雅俊教授送上崇高的敬意与万分的感激.

由于我们水平有限,书中难免存在这样或者那样的问题,恳请各位同行和读者批评指正. 我们期待本书能不断完善,也期待有更优秀的教材面世. 让我们共同努力在 ISEC 平台上成长壮大,为适应国际化的教育发展做好充分的准备.

作 者

2024 年 1 月修订

目 录

Chapter 1　Preliminary Analysis of Space Analytic Geometry
第 1 章　空间解析几何初步 …………… 1

1.1　Vectors and Linear Operations
1.1　向量及线性运算 …………… 1
　　1.　The Concept of Vector
　　1.　向量的概念 …………… 1
　　2.　Linear Operations of Vectors
　　2.　向量的线性运算 …………… 4
　　3.　Space Cartesian Coordinate System
　　3.　空间直角坐标系 …………… 7

1.2　Scalar Product and Cross Product
1.2　数量积与向量积 …………… 13
　　1.　Definition and Operation Law of Scalar Product
　　1.　数量积的定义及运算规律 …………… 13
　　2.　Cartesian Coordinate Operation of Scalar Product
　　2.　数量积的直角坐标运算 …………… 14
　　3.　The Definition and Operation Rule of Cross Product
　　3.　向量积的定义及运算规律 …………… 15
　　4.　Cartesian Coordinate Operation of Cross Product
　　4.　向量积的直角坐标运算 …………… 15
　　5.　The Relationship and Its Judgement of Vectors
　　5.　向量的关系及其判定 …………… 16

1.3　Plane and Its Equation
1.3　平面及其方程 …………… 19
　　1.　Point Normal form Equation of the Plane
　　1.　平面的点法式方程 …………… 19
　　2.　General Equation of the Plane
　　2.　平面的一般式方程 …………… 20
　　3.　Intercept Equation of the Plane
　　3.　平面的截距式方程 …………… 21
　　4.　Three Points Equation of the Plane
　　4.　平面的三点式方程 …………… 21
　　5.　The Angle Between Two Planes and the Positional Relationship
　　5.　两平面的夹角和位置关系 …………… 22
　　6.　Distance from Point to Plane
　　6.　点到平面的距离 …………… 23

1.4　Space Straight Lines and Their Equations
1.4　空间直线及其方程 …………… 24
　　1.　Symmetric Equation of a Straight Line
　　1.　直线的对称式方程 …………… 24
　　2.　Parametric Equation of a Straight Line
　　2.　直线的参数式方程 …………… 26
　　3.　General Equation of a Straight Line
　　3.　直线的一般式方程 …………… 26
　　4.　The General Formula of Linear Equation and Transformation of Symmetric Formula
　　4.　直线方程的一般式与对称式的转化 …………… 26
　　5.　The Angle and Positional Relation Between Two Straight Lines in Space
　　5.　空间中两直线的夹角和位置关系 …………… 27
　　6.　The Angle and Position Relation Between a Line and a Plane
　　6.　直线与平面的夹角和位置关系 …… 28
　　7.　Distance from Point to Line
　　7.　点到直线的距离 …………… 29

1.5 Quadratic Surfaces and Their Equations
1.5 二次曲面及其方程 ………… 31
   1. Spherical Surface
   1. 球面 ………… 31
   2. Ellipsoid
   2. 椭球面 ………… 32
   3. Hyperboloid
   3. 双曲面 ………… 33
   4. Paraboloid
   4. 抛物面 ………… 35
   5. Cylinder
   5. 柱面 ………… 36
   6. Rotating Surface
   6. 旋转曲面 ………… 37
1.6 Space Curves and Their Equations
1.6 空间曲线及其方程 ………… 41
   1. General Equation of Space Curve
   1. 空间曲线的一般方程 ………… 41
   2. Parametric Equation of Space Curve
   2. 空间曲线的参数方程 ………… 42
   3. The Projection of a Space Curve on a Coordinate Surface
   3. 空间曲线在坐标面上的投影 ………… 43
Exercises 1
习题 1 ………… 48

Chapter 2 Derivatives for the Function of Several Variables
第 2 章 多元函数的微分 ………… **54**
2.1 The Basic Concept of the Function of Several Variables
2.1 多元函数的基本概念 ………… 54
   1. Planar Point Set
   1. 平面点集 ………… 54
   2. The Concept of the Function of Several Variables
   2. 多元函数的概念 ………… 57
2.2 Limit and Continuity of the Function of Two Variables
2.2 多元函数的极限与连续性 ………… 60
   1. Limit of the Function of Two Variables
   1. 二元函数的极限 ………… 60
   2. Continuity of the Function of Two Variables
   2. 二元函数的连续性 ………… 65
2.3 Partial Derivatives
2.3 偏导数 ………… 68
   1. Concept of the Partial Derivatives
   1. 偏导数的概念 ………… 68
   2. Rule for Finding Partial Derivatives
   2. 求偏导数的法则 ………… 69
   3. Geometric Interpretations of Partial Derivative
   3. 偏导数的几何解释 ………… 71
   4. Partial Derivatives of Higher Order
   4. 高阶偏导数 ………… 73
   5. More than Two Variables
   5. 多于两个变量的情形 ………… 74
2.4 Total Differential
2.4 全微分 ………… 76
   1. The Concept of Total Differential
   1. 全微分的概念 ………… 76
   2. The Application of Total Differential in Approximate Calculation
   2. 全微分在近似计算中的应用 ………… 78
2.5 The Derivative Rule of Multivariate Composite Function
2.5 多元复合函数的求导法则 ………… 79
2.6 The Derivative Rule of Implicit Function
2.6 隐函数的求导法则 ………… 83
2.7 Local Extremum, Maximum and Minimum
2.7 局部极值,最值 ………… 84
   1. Local Extremum
   1. 局部极值 ………… 85
   2. Maximum and Minimum
   2. 最值 ………… 88
Exercises 2
习题 2 ………… 91

Chapter 3 Double Integral
第3章 二重积分 ·············· **97**

- 3.1 The Double Integral on Closed Rectangles
- 3.1 闭矩形区域上的二重积分 ·········· 97
  - 1. The Definition of the Double Integral
  - 1. 二重积分的定义 ············ 98
  - 2. The Existence Question of Double Integral
  - 2. 二重积分的存在性问题 ········· 101
  - 3. Properties of the Double Integral
  - 3. 二重积分的性质 ············ 102
  - 4. Simple Calculation of Double Integrals
  - 4. 二重积分的简单计算 ·········· 103
- 3.2 Iterated Integrals
- 3.2 累次积分 ··············· 106
  - 1. Change the Double Integral to the Iterated Integral
  - 1. 化二重积分为累次积分 ········· 106
  - 2. Calculating Iterated Integral
  - 2. 累次积分的计算 ············ 108
- 3.3 The Double Integral on Non Closed Rectangular Regions
- 3.3 非闭矩形区域上的二重积分 ········ 111
  - 1. The Definition of Double Integral on a Bounded Closed Area
  - 1. 有界闭区域上的二重积分的定义 ············· 111
  - 2. Calculation of Double Integral on a Bounded Closed Area
  - 2. 有界闭区域上的二重积分的计算 ············· 112
- 3.4 The Double Integral in Polar Coordinates
- 3.4 极坐标下的二重积分 ··········· 120
- 3.5 Applications of Double Integral
- 3.5 二重积分的应用 ············· 128
  - 1. The Quality of Flat Sheet
  - 1. 平面薄板的质量 ············ 128
  - 2. Center of Mass of Flat Sheet
  - 2. 平面薄板的质心 ············ 130
  - 3. The Moment of Inertia of a Flat Sheet
  - 3. 平面薄板的转动惯量 ·········· 132
- Exercises 3
- 习题 3 ··················· 133

Chapter 4 Infinite Series
第4章 无穷级数 ·············· **138**

- 4.1 Determine Whether the Infinite Series Converges or Diverges
- 4.1 判断无穷级数的敛散性 ············ 139
  - 1. The Concept of Convergence and Divergence of Series
  - 1. 级数敛散性的概念 ··········· 139
  - 2. The Basic Property of the Series
  - 2. 级数的基本性质 ············ 142
- 4.2 The Positive Terms Series
- 4.2 正项级数 ················ 144
- 4.3 Alternating Series, Absolute Convergence and Conditional Convergence
- 4.3 交错级数，绝对收敛和条件收敛 ················ 150
  - 1. Alternating Series and Its Tests for Convergence
  - 1. 交错级数及其收敛判别法 ······ 150
  - 2. Absolute and Conditional Convergence
  - 2. 绝对收敛与条件收敛 ·········· 152
- 4.4 Power Series
- 4.4 幂级数 ················· 154
- 4.5 Operations and Properties of Power Series
- 4.5 幂级数的运算与性质 ··········· 160
  - 1. Operations of Power Series
  - 1. 幂级数的运算 ············· 160
  - 2. Properties of Power Series
  - 2. 幂级数的性质 ············· 160
- Exercises 4
- 习题 4 ··················· 164

Chapter 1　Preliminary Analysis of Space Analytic Geometry
第 1 章　空间解析几何初步

In plane analytic geometry, with coordinate system a point on the plane can be corresponded to a binary ordered array (x,y), and the graphs can be corresponded to their equations, which enables geometric problems to be solved with algebraic methods. Similarly, the points in the three-dimensional space can also be corresponded to the ternary ordered array (x,y,z), and this makes it possible for us to study space geometry in the same way.

This chapter first introduces the concepts concerning the vector as well as rules involved in its operation. Then by establishing a Cartesian coordinate system we move on to the study of the plane, space, lines, several curves and curved surfaces with algebraic methods. All those will contribute to the further study of multivariate functional calculus.

平面解析几何中,通过坐标法把平面上的点与二元有序数组(x,y)对应起来,把平面图形和方程对应起来,从而可以用代数方法来研究几何问题.类似地,空间解析几何将空间中的点与三元有序数组(x,y,z)对应起来,进而用代数方法来研究空间几何问题.

本章首先引进向量的概念及运算,进而建立空间直角坐标系,然后利用代数方法研究空间平面和直线以及几种特殊的曲面和曲线.这些内容对学习多元函数微积分将起到重要的作用.

1.1　Vectors and Linear Operations
1.1　向量及线性运算

1. The Concept of Vector

Time, mass, work, length, area and volume are all measurements only with magnitude, and they are called **quantities**. There are, however, other measurements such as displacement, force, velocity and acceleration, which has not only magnitude but also direction, and they are called vectors.

Definition 1.1　Quantity that have both magnitude and direction are called **vector**.

Vectors are usually expressed as directed line segments, whose starting point and end point are also the **starting point** and **end point** of the vectors. The direction of the directed line segment is the direction of the vector,

1. 向量的概念

我们经常遇到的像时间、质量、功、长度、面积与体积等这种只有大小的量叫作**数量**.像位移、力、速度、加速度等这种不但有大小,而且还有方向的量就是向量.

定义 1.1　既有大小又有方向的量叫作**向量**或**矢量**.

我们通常用有向线段表示向量:有向线段的始点与终点分别叫作向量的**始点**和**终点**,有向线段的方向表示向量的方向,而有向线段的长度代表向量的大小.例如,始点是

and its length can be measured as the magnitude of the vector. For example, a vector started with A and ended with B can be written as $\overrightarrow{AB}$(Figure 1.1(a)). When written vectors can be recorded as lowercase letters with an arrow above, such as $\vec{a}, \vec{b}, \vec{c}, \vec{l}$. When printed they can be expressed as boldface letters, such as $\boldsymbol{a}, \boldsymbol{b}, \boldsymbol{c}, \boldsymbol{l}$ (Figure 1.1(b)).

A,终点是 B 的向量记作 $\overrightarrow{AB}$(图 1.1(a)). 在手写时,常常用带箭头的小写字母来表示向量,如 $\vec{a}, \vec{b}, \vec{c}, \vec{l}$ 等. 而在印刷时,常常用黑体字母来记向量,如 $\boldsymbol{a}, \boldsymbol{b}, \boldsymbol{c}, \boldsymbol{l}$ 等(图 1.1(b)).

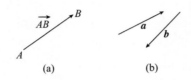

Figure 1.1

图 1.1

The magnitude of a vector is called the **modulus** of a vector, also called the **length** of a vector. Vector $\overrightarrow{AB}$ and the modulus of $\boldsymbol{a}$ are recorded separately as $|\overrightarrow{AB}|$ and $|\boldsymbol{a}|$.

A vector with module 1 is called a **unit vector**. The unit vector in the same direction as vector $\boldsymbol{a}$ is called the unit vector of $\boldsymbol{a}$. It is often represented by $\boldsymbol{a}^0$.

A vector with module 0 is called a **zero vector** and is denoted as **0**. It is the vector where the starting point coincides with the destination. The direction of the zero vector is uncertain and may be in any direction.

Given that vectors are treated as directed line segment in geometry. Similarly, a vector can also be parallel to a line or a plane. So as with line segments, the following says that vector $\boldsymbol{a}$ is **parallel** to $\boldsymbol{b}$, meaning that their lines are parallel to each other and remember to be $\boldsymbol{a}//\boldsymbol{b}$. Similarly, we can also be parallel to a straight line or a plane.

Definition 1.2 If the two vector modules are equal and have the same direction, then these two vectors are called **equal** vectors. If vector $\boldsymbol{a}$ is equal to $\boldsymbol{b}$, then it's recorded as $\boldsymbol{a}=\boldsymbol{b}$.

All zero vectors are equal.

According to Definition 1.2, for two nonzero vectors $\overrightarrow{AB}$ and $\overrightarrow{A'B'}$ which equal to each other and are non-colinear,

向量的大小叫作向量的**模**,也称为向量的**长度**,向量 $\overrightarrow{AB}$ 和 $\boldsymbol{a}$ 的模分别记作 $|\overrightarrow{AB}|$ 和 $|\boldsymbol{a}|$.

模等于 1 的向量叫作**单位向量**;与向量 $\boldsymbol{a}$ 具有同一方向的单位向量叫作向量 $\boldsymbol{a}$ 的单位向量,常常用 $\boldsymbol{a}^0$ 来表示.

模等于 0 的向量叫作**零向量**,记作 **0**,它是起点与终点重合的向量. 零向量的方向不确定,可以是任意方向.

由于在几何中我们把向量看成有向线段,因此像对待线段一样,下面说到向量 $\boldsymbol{a}$ 与 $\boldsymbol{b}$ 平行,意思就是它们所在的直线平行,并记作 $\boldsymbol{a}//\boldsymbol{b}$. 类似地,我们可以说一个向量与一条直线或一个平面平行等.

定义 1.2 如果两个向量的模相等且方向相同,那么称这两个向量**相等**. 向量 $\boldsymbol{a}$ 与 $\boldsymbol{b}$ 相等,记作 $\boldsymbol{a}=\boldsymbol{b}$.

所有的零向量都相等.

根据定义 1.2,对于不在同一直线上的两个相等的非零向量 $\overrightarrow{AB}$ 与 $\overrightarrow{A'B'}$,如果用线

connect their starting points A and A' as well as end points B and B', and you will get a parallelogram $ABB'A'$ (Figure 1.2). Conversely, two vectors can be said to be equal if they can form a parallelogram in the same way.

段分别连接它们的一对起点 A 与 A' 和一对终点 B 与 B',那么显然得到一个平行四边形 $ABB'A'$(图 1.2). 反过来,如果用这种做法从两个向量得到一个平行四边形,那么这两个向量就相等.

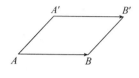

Figure 1.2
图 1.2

Whether two vectors are equal has nothing to do with their starting points, but it depends on their moduli and directions. These are also the vectors we will use in our study, **free vectors**, whose modulus and directions matter, but their starting points are worth no consideration. In other words, their movement will not change what they are. Given the arbitrariness of the starting points, we may pick certain point as the common starting point shared by several vectors to facilitate our study.

It must be noted that since vectors are not only of magnitude, but also has direction, the vectors of equal modulus are not necessarily equal because their directions cannot be the same.

Definition 1.3 Two vectors whose moduli are equal but directions are opposite are **reverse vectors**. The opposite of the vector a is denoted as $-a$.

Obviously, vector $\overrightarrow{AB}$ and $\overrightarrow{BA}$ are reverse vectors, that is
$$\overrightarrow{AB}=-\overrightarrow{BA}, \quad \text{or} \quad \overrightarrow{BA}=-\overrightarrow{AB}.$$

Definition 1.4 There are nonzero vectors a, b. From the same starting point O, the directed line segments $\overrightarrow{OA}$ and $\overrightarrow{OB}$ represent a and b respectively. The angle formed by the rays OA and OB whose angles between 0 and π becomes the **angle** between a and b, written as $\langle a, b \rangle$.

两个向量是否相等与它们的始点无关,只由它们的模和方向决定. 我们以后运用的正是这种始点可以任意选取,而只由模和方向决定的向量. 这样的向量通常叫作**自由向量**. 也就是说,自由向量可以任意平行移动,移动后的向量仍代表原来的向量. 由于自由向量始点的任意性,我们可以按需要选取某一点作为所研究的一些向量的公共始点.

必须注意,由于向量不仅有大小,而且还有方向,因此模相等的两个向量不一定相等,因为它们的方向可能不同.

定义 1.3 两个模相等、方向相反的向量互为**反向量**. 向量 a 的反向量记作 $-a$.

显然,向量 $\overrightarrow{AB}$ 与 $\overrightarrow{BA}$ 互为反向量,即
$$\overrightarrow{AB}=-\overrightarrow{BA} \quad \text{或} \quad \overrightarrow{BA}=-\overrightarrow{AB}.$$

定义 1.4 设有非零向量 a,b,从同一始点 O 作有向线段 $\overrightarrow{OA}, \overrightarrow{OB}$ 分别表示 a 与 b. 由射线 OA 和 OB 构成的角度在 0 与 π 之间的角称为 a 与 b 的**夹角**,记作 $\langle a,b \rangle$.

In particular, if $\langle a,b\rangle=\dfrac{\pi}{2}$, the vectors a and b are **vertical**, written as $a\perp b$. Zero vector can be said to be vertical to any other vector.

Definition 1.5 A group of vectors parallel to the same line are called **collinear vectors**. The zero vector is collinear with any collinear vector group.

Definition 1.6 A group of vectors parallel to the same plane are called **coplanar vectors**. Zero vectors are coplanar with any vectors.

Obviously, a set of collinear vectors must be coplanar vectors, and if the two vectors are collinear in the three vectors, the three vectors must be coplanar.

2. Linear Operations of Vectors

1) Addition of Vectors

Definition 1.7 The nonzero vector a, b are represented by directed line segments $\overrightarrow{OA}$, $\overrightarrow{OB}$, respectively, which are from the same starting point O. We make a parallelogram $OACB$ with the adjacent sides $\overrightarrow{OA}$, $\overrightarrow{OB}$. We call the diagonal vector $\overrightarrow{OC}$ of the parallelogram as the **sum** of a and b, recorded as $a+b$, that is
$$\overrightarrow{OC}=a+b=\overrightarrow{OA}+\overrightarrow{OB}.$$

The vector addition rules given by Definition 1.7 is called the **parallelogram rule** (Figure 1.3). If we let the end point of vector a be the starting point of vector b, then the vector from the starting point of a to the end point of b is also the sum of a and b, that is $a+b$ (Figure 1.4). In fact, we just shift the vector $\overrightarrow{OB}$ of Figure 1.3 to the position of $\overrightarrow{AC}$. This method is called the **Triangle Rule**.

特别地,若$\langle a,b\rangle=\dfrac{\pi}{2}$,则称向量$a$与$b$**垂直**,记作$a\perp b$.可以认为零向量与任何向量垂直.

定义1.5 平行于同一直线的一组向量叫作**共线向量**.零向量与任何共线的向量组共线.

定义1.6 平行于同一平面的一组向量叫作**共面向量**.零向量与任何共面的向量组共面.

显然,一组共线向量一定是共面向量;三向量中,如果有两个向量是共线的,则这三个向量一定也是共面的.

2. 向量的线性运算

1) 向量的加法

定义1.7 设有非零向量a,b.从同一始点O作有向线段$\overrightarrow{OA},\overrightarrow{OB}$分别表示$a$与$b$,然后以$\overrightarrow{OA},\overrightarrow{OB}$为邻边作平行四边形$OACB$.我们把平行四边形的对角线向量$\overrightarrow{OC}$称为向量$a$与$b$的和(图1.3),记作$a+b$,即
$$\overrightarrow{OC}=a+b=\overrightarrow{OA}+\overrightarrow{OB}.$$

定义1.7给出的向量加法规则称为**平行四边形法则**(图1.3).如果以向量a的终点作为向量b的始点,则由a的始点到b的终点的向量就是a与b的和$a+b$(图1.4).事实上,只要将图1.3中的向量$\overrightarrow{OB}$平移到$\overrightarrow{AC}$的位置就行了.这种求向量和的方法称为**三角形法则**.

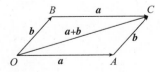

Figure 1.3
图1.3

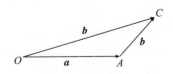

Figure 1.4
图1.4

Clearly, we have
$$a+0=a,$$
$$a+(-a)=0.$$

The sum of the two vectors **a** and **b** is called **vector addition**. The addition of vectors satisfies the following laws of operation:

(1) **Commutative law**: $a+b=b+a$;

(2) **Associative law**: $(a+b)+c=a+(b+c)$.

Since the addition of vectors satisfies commutative law and associative law, the addition of n vectors $a_1, a_2, \cdots, a_n$ $(n \geqslant 3)$ could be written as
$$a_1+a_2+\cdots+a_n.$$

2) Vector Subtraction

Definition 1.8 When the sum of the vector **b** and the vector **c** is equal to the vector **a**, that is, $b+c=a$, we call the vector **c** the **difference** between **a** and **b**, and recorded as
$$c=a-b.$$

The operation of calculating their subtraction is called **vector subtraction**.

Based on the Triangle Rule of vector addition, we have
$$\overrightarrow{OB}+\overrightarrow{BA}=\overrightarrow{OA}.$$
So, by Definition 1.8,
$$\overrightarrow{BA}=\overrightarrow{OA}-\overrightarrow{OB}.$$

Thus the geometric construction method of vector subtraction is obtained: suppose vectors $\overrightarrow{OA}=a, \overrightarrow{OB}=b$, and they all start with the same point O whose location is arbitrary in the coordinate system. Then vector $\overrightarrow{BA}=a-b$ (Figure 1.5). If $\overrightarrow{OA}$ and $\overrightarrow{OB}$ are a pair of adjacent edges that make up the parallelogram $OACB$, then the diagonal vector $\overrightarrow{OC}=a+b$, and another diagonal vector $\overrightarrow{BA}=a-b$ (Figure 1.6).

显然,有
$$a+0=a,$$
$$a+(-a)=0.$$

求两向量 **a** 与 **b** 的和的运算叫作**向量加法**. 容易得到向量的加法满足下面的运算规律:

(1) **交换律**: $a+b=b+a$;

(2) **结合律**: $(a+b)+c=a+(b+c)$.

由于向量的加法满足交换律和结合律,因此 n 个向量 $a_1, a_2, \cdots, a_n (n \geqslant 3)$ 相加可写成
$$a_1+a_2+\cdots+a_n.$$

2) 向量减法

定义 1.8 当向量 **b** 与向量 **c** 的和等于向量 **a**,即 $b+c=a$ 时,我们把向量 **c** 叫作 **a** 与 **b** 的**差**,并记作
$$c=a-b.$$

由两向量 a, b 求它们的差的运算叫作**向量减法**.

根据向量加法的三角形法则,有
$$\overrightarrow{OB}+\overrightarrow{BA}=\overrightarrow{OA},$$
所以由定义 1.8 得
$$\overrightarrow{BA}=\overrightarrow{OA}-\overrightarrow{OB}.$$

由此得到向量减法的几何作图法:自空间任意点 O 引向量 $\overrightarrow{OA}=a, \overrightarrow{OB}=b$,那么向量 $\overrightarrow{BA}=a-b$(图 1.5). 如果以 $\overrightarrow{OA}, \overrightarrow{OB}$ 为一对邻边作平行四边形 $OACB$,那么显然它的一条对角线向量为 $\overrightarrow{OC}=a+b$,而另一条对角线向量为 $\overrightarrow{BA}=a-b$(图 1.6).

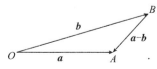

Figure 1.5

图 1.5

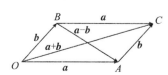

Figure 1.6

图 1.6

By using the opposite vector, vector subtraction can be converted to addition:
$$a-b=a+(-b).$$
This indicates that $a-b$ is the sum of a and $-b$ which is the inverse vector of b. Because b is the inverse vector of $-b$, we can obtain that
$$a-(-b)=a+b.$$

We can obtain the transportation rule of vectors from the characters of vector subtraction. That is, if we want to put a vector on the other side of the equality sign, we only need to transform it into its opposite. For example, in the equation $a+b+c=d$, if we want to put c on the right side of the equality sign, we will get $a+b=d-c$. This is because the transformation needs a c deducted from both sides, that is adding $-c$, and $c+(-c)=0$.

3) Scalar Multiplication

We understand n (the number of vectors) vectors added together still produce a vector, especially in the case where n vectors, a, are added. In that case, the modulus of the sum vector $|a|$ is n times of a, and the direction is the same as a. The operation can be recorded as na.

Definition 1.9 The product of real number λ and the vector a is a vector, recorded as λa, its modulus is
$$|\lambda a|=|\lambda||a|.$$
The direction of λa: when $\lambda>0$, λa and a are identical; when $\lambda<0$, λa and a are reverse; when $\lambda=0$, $\lambda a=0$.

We call this operation the **multiplication of numbers and vectors**, referred to as **multiplication**.

For arbitrary vectors a, b and real number λ, μ, the following algorithms are used, that is, the multiplication of numbers and vectors satisfies the following laws of operation:

(1) $(\lambda\mu)a=\lambda(\mu a)$;

(2) $(\lambda+\mu)a=\lambda a+\mu a$;

(3) $\lambda(a+b)=\lambda a+\lambda b$.

Specifically, when $\lambda=-1$, $(-1)a$ is the opposite vector of a, therefore we often abbreviate $(-1)a$ to $-a$.

Given the vector a and its unit vector a^0, the following

利用反向量,可以把向量减法运算变为加法运算:
$$a-b=a+(-b).$$
这表明 $a-b$ 为求 a 与 b 的反向量 $-b$ 之和. 又因为 $-b$ 的反向量就是 b,因此又可得
$$a-(-b)=a+b.$$

从向量减法的这个性质,可以得出向量等式的移向法则:在向量等式中,将某一向量从等号的一端移到另一端,只需要改变它的符号. 例如,将等式 $a+b+c=d$ 中的 c 移到另一端,那么有 $a+b=d-c$. 这是因为从等式 $a+b+c=d$ 两边减去 c,即加上 $-c$,而 $c+(-c)=0$ 的缘故.

3) 数乘向量

我们知道,在向量的加法中,n 个向量相加仍然是向量,特别是 n 个相同的非零向量 a 相加的情形,显然这时和向量的模为 $|a|$ 的 n 倍,方向与 a 相同. n 个 a 相加的和记作 na.

定义 1.9 实数 λ 与向量 a 的乘积是一个向量,记作 λa,其模是
$$|\lambda a|=|\lambda||a|,$$
其方向是:当 $\lambda>0$ 时,λa 与 a 同向;当 $\lambda<0$ 时,λa 与 a 反向;当 $\lambda=0$ 时,$\lambda a=0$,它的方向为任意方向.

我们把这种运算叫作**数量与向量的乘法**,简称**数乘**.

对于任意向量 a,b 以及任意实数 λ,μ,有以下运算法则,即数量与向量的乘法满足下面的运算规律:

(1) $(\lambda\mu)a=\lambda(\mu a)$;

(2) $(\lambda+\mu)a=\lambda a+\mu a$;

(3) $\lambda(a+b)=\lambda a+\lambda b$.

特别地,当 $\lambda=-1$ 时,$(-1)a$ 就是 a 的反向量,因此我们常常把 $(-1)a$ 简写成 $-a$.

对于非零向量 a 和它的单位向量 a^0,下

equation is established:

$$a = |a|a^0 \quad \text{or} \quad a^0 = \frac{a}{|a|}.$$

Accordingly, a nonzero vector is multiplied by the reciprocal of its modulus, and the result is a unit vector in the same direction as it.

The addition, subtraction, and multiplication of vectors are called **linear operations of vectors**, and $\lambda a + \mu b$ is called a linear combination of a, b ($\lambda, \mu \in \mathbf{R}$).

3. Space Cartesian Coordinate System

We need a coordinate system to determine the location of one point in the space.

1) Space Cartesian Coordinate System

Go through a point O in space, and use three of mutually perpendicular axes, x-axis, y-axis, and z-axis, so as to form the **space Cartesian coordinate system**, which is recorded as $Oxyz$, where the general provisions of the x-axis, y-axis, z-axis position follow right-hand rule: let the four fingers of the right hand the positive direction of point to the x-axis, and four fingers along the direction of the positive y-axis of the fist, then the direction of the thumb is positive. Usually in all axes the length of a unit are the same. Place the x-axis and y-axis on the horizontal plane and the z-axis is perpendicular to the horizontal plane(Figure 1.7).

In the space coordinate system $Oxyz$, the point O is called **coordinate origin**, which is called the **origin**. x-axis, y-axis, z-axis, are known as the **horizontal axis**, **ordinate axis**, **vertical axis**, and they are collectively known as **coordinate axis**. Any two of the three axes can form a **coordinate plane**. There are all together three coordinate planes, and they are Oxy, Oyz and Oxz, which divide the space into 8 parts. They are called the 1st, 2nd, 3rd, 4th, 5th, 6th, 7th, and 8th octant respectively. The first octant is located above the surface formed by the positive segment of x-axis and y-axis. The 2nd, 3rd and 4th octants are also above the Oxy surface and they are arranged in the anticlockwise order. The 5th octant is right below the first octant, and the 6th, 7th and

面的等式成立:

$$a = |a|a^0 \quad 或 \quad a^0 = \frac{a}{|a|}.$$

由此可知,一个非零向量乘以它的模的倒数,结果是一个与它同方向的单位向量.

向量的加法、减法及数乘运算统称为**向量的线性运算**;$\lambda a + \mu b$ 称为 a, b 的一个线性组合($\lambda, \mu \in \mathbf{R}$).

3. 空间直角坐标系

若想确定空间中一点的位置,就需要建立坐标系.

1) 空间直角坐标系

过空间中一点 O,作三条两两互相垂直的数轴,分别标记为 x 轴,y 轴和 z 轴,这样就构成了**空间直角坐标系**,记作 $Oxyz$,其中一般规定 x 轴,y 轴和 z 轴的位置关系遵循右手法则:让右手的四个手指指向 x 轴的正向,然后让四指沿握拳方向转向 y 轴的正向,大拇指所指的方向为 z 轴的正向.通常在各数轴上的单位长度相同;把 x 轴,y 轴放置在水平平面上,让 z 轴垂直于水平平面(图 1.7).

在空间直角坐标系 $Oxyz$ 中,点 O 称为**坐标原点**,简称原点.x 轴,y 轴,z 轴这三条数轴,分别称为**横轴**、**纵轴**、**竖轴**,统称为**坐标轴**.由任意两条坐标轴所确定的平面称为**坐标面**,共有 Oxy,Oyz,Ozx 三个坐标面.三个坐标面把空间分隔成八个部分,每个部分分别称为第一、第二……第八卦限,其中第一卦限位于 x 轴,y 轴,z 轴的正向位置,第二至四卦限也位于 Oxy 坐标面的上方,按逆时针方向排列;第五卦限在第一卦限的正下方,第六至八卦限也在 Oxy 坐标面的下方,按逆时针方向排列(图 1.8).

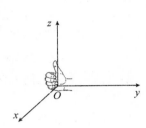

Figure 1.7
图 1.7

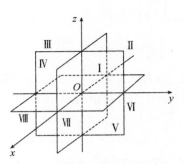

Figure 1.8
图 1.8

8th octants are also arranged in the anticlockwise order(Figure 1.8).

2) Cartesian Coordinates of Spatial Points

We will establish the relation between the point in space and the ordered array composed of three real numbers, namely the relation between the midpoint of the space and the coordinate.

2) 空间点的直角坐标

我们将通过空间直角坐标系,建立空间中的点与由三个实数组成的有序数组的关系,即空间中的点与坐标之间的关系.

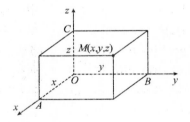

Figure 1.9
图 1.9

As shown in Figure 1.9, let M be any point in space. Passing the point M, we make three planes perpendicular to the three coordinate axes, respectively, and these three planes intersect with x-axis, y-axis and z-axis at three points A,B,C. If the coordinates of these three points in x-axis, y-axis, z-axis are respectively x,y,z. M uniquely determines a ternary ordered array (x,y,z), which is called the coordinate of M in space Cartesian coordinate system and denoted as $M(x,y,z)$. x,y,z are called x component, y component, z component of the coordinate of M. On the other hand, if a ternary ordered array (x,y,z) is given arbitrarily, we can only determine a point in space Cartesian coordinate system. In fact, the points A,B,C with the coordinates x,y,z are sequentially

如图 1.9 所示,设 M 为空间中的任意一点,过点 M 分别作垂直于三个坐标轴的三个平面,与 x 轴,y 轴和 z 轴依次交于 A,B,C 三点. 若这三点在 x 轴,y 轴,z 轴上的坐标分别为 x,y,z,点 M 就唯一地确定了一组三元有序数组 (x,y,z). 称该三元有序数组 (x,y,z) 为点 M 在空间直角坐标系中的坐标,记作 $M(x,y,z)$,其中 x,y,z 分别称为点 M 坐标的 x,y 和 z 分量. 反之,如果任给一组三元有序数组 (x,y,z),在空间直角坐标系中可唯一确定一点. 事实上,依次在 x 轴,y 轴,z

taken on the x-axis, y-axis and z-axis. Make a plane at each point so that these three planes are perpendicular to x-axis, y-axis and z-axis. The intersection of the three planes is the only point determined by (x,y,z). In this way, the point in space and the ternary ordered array establish one-to-one correspondence.

The coordinates of origin O are 0, that is $O(0,0,0)$. If the point M is on the Oxy coordinate plane, the coordinate of M is $(x,y,0)$; if the point M is on the x-axis, the coordinate of M is $(x,0,0)$. We can similarly get to the coordinates of points on other coordinate surfaces or axes. The three coordinates of the eight interior points of the hexagram are not zero, and the symbols of each component are determined by the point where the hexagram is located.

Considering the rectangular coordinate system, we can dig into the relationship of two points symmetrical from the axis, the coordinate plane or the origin. For example, the symmetry point of (x,y,z) from x-axis is $(x,-y,-z)$; the symmetry point of (x,y,z) from Oxy-plane is $(x,y,-z)$; the symmetry point of (x,y,z) from the origin is $(-x,-y,-z)$ and so on.

3) Coordinate Representation of Vectors

In the space Cartesian coordinate system $Oxyz$, on the x-axis, the y-axis and the z-axis, each unit vector corresponding to the coordinate axis is taken as i, j, k, which are called **coordinate vectors**. Any vector a in space can be uniquely represented by the sum of i, j, k.

In fact, let $a = \overrightarrow{MN}$, passing the points M, N, we make a plane which parallel to the coordinate plane. We could get a rectangular parallelopiped with the diagonal MN, as shown in Figure 1.10, we have
$$a = \overrightarrow{MN} = \overrightarrow{MA} + \overrightarrow{AP} + \overrightarrow{PN}$$
$$= \overrightarrow{MA} + \overrightarrow{MB} + \overrightarrow{MC}.$$

Since $\overrightarrow{MA}$ is parallel to i, $\overrightarrow{MB}$ is parallel to j, $\overrightarrow{MC}$ is parallel to k, there exists unique real numbers x, y, z, which make
$$\overrightarrow{MA} = xi, \quad \overrightarrow{MB} = yj, \quad \overrightarrow{MC} = zk,$$
that is
$$a = xi + yj + zk. \tag{1.1}$$

轴上取坐标为 x,y,z 的点 A,B,C,过这三点各作一平面,使其分别垂直于 x 轴,y 轴,z 轴,这三个平面的交点就是 (x,y,z) 确定的唯一点. 这样,空间中的点便与三元有序数组建立了一一对应的关系.

显然,原点 O 的各坐标分量均为 0,即 $O(0,0,0)$;若点 M 在 Oxy 坐标面上,则 M 的坐标为 $(x,y,0)$;若点 M 在 x 轴上,则 M 的坐标为 $(x,0,0)$. 类似可得其他坐标面或坐标轴上点的坐标特征. 八个卦限内点的三个坐标均不为零,各分量的符号由点所在卦限确定.

类似于平面直角坐标系的情形,可以讨论关于坐标轴、坐标面、坐标原点对称的点的坐标关系. 例如,点 (x,y,z) 关于 x 轴对称的点为 $(x,-y,-z)$;点 (x,y,z) 关于 Oxy 坐标面对称的点为 $(x,y,-z)$;点 (x,y,z) 关于原点对称的点为 $(-x,-y,-z)$;等等.

3) 向量的坐标表示

在空间直角坐标系 $Oxyz$ 中,在 x 轴,y 轴,z 轴上各取一个与坐标轴同向的单位向量,依次记作 i,j,k,它们称为**坐标向量**. 空间中任一向量 a 都可以唯一地表示为 i,j,k 的数乘之和.

事实上,设 $a = \overrightarrow{MN}$,过点 M,N 分别作平行于坐标平面的平面,得到一个以线段 MN 为对角线的长方体,如图 1.10 所示,则有
$$a = \overrightarrow{MN} = \overrightarrow{MA} + \overrightarrow{AP} + \overrightarrow{PN}$$
$$= \overrightarrow{MA} + \overrightarrow{MB} + \overrightarrow{MC}.$$

由于 $\overrightarrow{MA}$ 与 i 平行,$\overrightarrow{MB}$ 与 j 平行,$\overrightarrow{MC}$ 与 k 平行,所以存在唯一的实数 x,y,z,使得
$$\overrightarrow{MA} = xi, \quad \overrightarrow{MB} = yj, \quad \overrightarrow{MC} = zk,$$
即
$$a = xi + yj + zk. \tag{1.1}$$

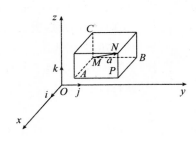

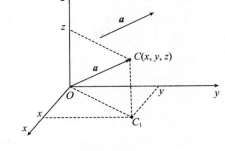

Figure 1.10
图 1.10

Figure 1.11
图 1.11

$x\boldsymbol{i}, y\boldsymbol{j}, z\boldsymbol{k}$ in equation (1.1) are known as the **component vectors** about $\boldsymbol{a}$ at x-axis, y-axis, z-axis, respectively. At the same time, an ordered array (x, y, z) consisted by the coefficients of $\boldsymbol{i}, \boldsymbol{j}, \boldsymbol{k}$ is called the **coordinate** of $\boldsymbol{a}$, recorded as
$$\boldsymbol{a} = \{x, y, z\}.$$
Coordinates vector is determined, vector can be determined.

Obviously, for the coordinate vector $\boldsymbol{i}, \boldsymbol{j}, \boldsymbol{k}$, there is
$$\boldsymbol{i} = \{1, 0, 0\}, \quad \boldsymbol{j} = \{0, 1, 0\}, \quad \boldsymbol{k} = \{0, 0, 1\}.$$

Suppose that the starting point of known vector $\boldsymbol{a}$ moved to the origin O, the end point is C, namely $\boldsymbol{a} = \overrightarrow{OC}$. $\overrightarrow{OC}$ is called the **diameter** (or **radius**). From the above formula (1.1) we know that the coordination (x, y, z) of point C is the coordinate of $\boldsymbol{a}$, the coordinate of vector $\boldsymbol{a}$ is equal to the diameter of the end point coordinates. Thus, in the establishment of the Cartesian coordinate space, there is a corresponding relationship among the vector, the radius, and the coordinate (Figure 1.11).

It is easy to see from Figure 1.11. If $\boldsymbol{a} = \{x, y, z\}$ and $\overrightarrow{OC} = \{x, y, z\}$, then in right triangle OC_1C, we have
$$|\boldsymbol{a}|^2 = |\overrightarrow{OC}|^2 = |OC_1|^2 + |C_1C|^2$$
$$= x^2 + y^2 + z^2,$$
that is
$$|\boldsymbol{a}| = \sqrt{x^2 + y^2 + z^2}.$$

4) Coordinate Formula of Linear Operations of Vectors

After introducing the coordinates of vectors, the operation of vectors can be transformed into algebraic operations.

我们把(1.1)式中的 $x\boldsymbol{i}, y\boldsymbol{j}, z\boldsymbol{k}$，分别称为向量 $\boldsymbol{a}$ 在 x 轴，y 轴，z 轴上的**分向量**；同时将 $\boldsymbol{i}, \boldsymbol{j}, \boldsymbol{k}$ 的系数组成的有序数组 (x, y, z) 叫作向量 $\boldsymbol{a}$ 的**坐标**，记为
$$\boldsymbol{a} = \{x, y, z\}.$$
向量的坐标确定了，向量也就确定了.

显然，对于坐标向量 $\boldsymbol{i}, \boldsymbol{j}, \boldsymbol{k}$，有
$$\boldsymbol{i} = \{1, 0, 0\}, \boldsymbol{j} = \{0, 1, 0\}, \boldsymbol{k} = \{0, 0, 1\}.$$

设把已知向量 $\boldsymbol{a}$ 的始点移到原点 O 时，其终点在 C，即 $\boldsymbol{a} = \overrightarrow{OC}$. 称 $\overrightarrow{OC}$ 为向径（或矢径）. 从上述(1.1)式的导出容易看出点 C 的坐标 (x, y, z) 就是 $\boldsymbol{a}$ 的坐标，即向量 $\boldsymbol{a}$ 的坐标就是与其相等的向径的终点坐标. 这样在建立了直角坐标系的空间中，向量、向径、坐标之间就有了一一对应的关系(图 1.11).

容易从图 1.11 看出，若 $\boldsymbol{a} = \{x, y, z\}$，即 $\overrightarrow{OC} = \{x, y, z\}$，则在直角三角形 OC_1C 中有
$$|\boldsymbol{a}|^2 = |\overrightarrow{OC}| = |OC_1|^2 + |C_1C|^2$$
$$= x^2 + y^2 + z^2,$$
即
$$|\boldsymbol{a}| = \sqrt{x^2 + y^2 + z^2}.$$

4) 向量的线性运算坐标公式

引入向量的坐标以后，就可将向量的运算转化为代数运算.

In space Cartesian coordinate system $Oxyz$, vectors $\boldsymbol{a} = \{x_1, y_1, z_1\}$ and $\boldsymbol{b} = \{x_2, y_2, z_2\}$, then by the definition of coordinate, we have

$$\boldsymbol{a} = x_1 \boldsymbol{i} + y_1 \boldsymbol{j} + z_1 \boldsymbol{k},$$
$$\boldsymbol{b} = x_2 \boldsymbol{i} + y_2 \boldsymbol{j} + z_2 \boldsymbol{k}.$$

Therefore

$$\boldsymbol{a} \pm \boldsymbol{b} = (x_1 \boldsymbol{i} + y_1 \boldsymbol{j} + z_1 \boldsymbol{k}) \pm (x_2 \boldsymbol{i} + y_2 \boldsymbol{j} + z_2 \boldsymbol{k})$$
$$= (x_1 \pm x_2) \boldsymbol{i} + (y_1 \pm y_2) \boldsymbol{j} + (z_1 \pm z_2) \boldsymbol{k},$$
$$\lambda \boldsymbol{a} = \lambda (x_1 \boldsymbol{i} + y_1 \boldsymbol{j} + z_1 \boldsymbol{k})$$
$$= (\lambda x_1) \boldsymbol{i} + (\lambda y_1) \boldsymbol{j} + (\lambda z_1) \boldsymbol{k} \quad (\lambda \in \mathbf{R}).$$

So

$$\boldsymbol{a} \pm \boldsymbol{b} = \{x_1 \pm x_2, y_1 \pm y_2, z_1 \pm z_2\},$$
$$\lambda \boldsymbol{a} = \{\lambda x_1, \lambda y_1, \lambda z_1\}.$$

That is, the sum (difference) of vectors is equal to the sum (difference) of their coordinates, the coordinate of $\lambda \boldsymbol{a}$ is equal to the number λ multiplied by the coordinate of $\boldsymbol{a}$.

Furthermore, the following conclusions can be drawn:

In Figure 1.10, if we let the coordinates of two points M, N in space as $M(x_1, y_1, z_1), N(x_2, y_2, z_2)$, then

(1) **Vector Coordinate**:

$$\overrightarrow{MN} = \overrightarrow{ON} - \overrightarrow{OM}$$
$$= \{x_2, y_2, z_2\} - \{x_1, y_1, z_1\}$$
$$= \{x_2 - x_1, y_2 - y_1, z_2 - z_1\},$$

that is, the vector coordinate can be produced by the coordinate of the starting point deducted that of the end point.

(2) **Modulus of a Vector**:

$$|\overrightarrow{MN}| = \sqrt{(x_2 - x_1)^2 + (y_2 - y_1)^2 + (z_2 - z_1)^2}.$$

5) Distance Between Two Points in Space

Let $M(x_1, y_1, z_1), N(x_2, y_2, z_2)$ be two points in space, and then the formula of distance between two points M and N is obtained:

$$d = \overrightarrow{MN}$$
$$= \sqrt{(x_2 - x_1)^2 + (y_2 - y_1)^2 + (z_2 - z_1)^2}. \qquad (1.2)$$

6) Direction Angle and Cosine of a Vector

The angles between the nonzero vector $\boldsymbol{a}$ and positive of the three axes are α, β, γ, which are called the **direction angles** of the vector $\boldsymbol{a}$. The cosine of the direction angles

设在空间直角坐标系 $Oxyz$ 中,向量 $\boldsymbol{a} = \{x_1, y_1, z_1\}$ 及 $\boldsymbol{b} = \{x_2, y_2, z_2\}$,则由向量的坐标定义有

$$\boldsymbol{a} = x_1 \boldsymbol{i} + y_1 \boldsymbol{j} + z_1 \boldsymbol{k},$$
$$\boldsymbol{b} = x_2 \boldsymbol{i} + y_2 \boldsymbol{j} + z_2 \boldsymbol{k}.$$

因此

$$\boldsymbol{a} \pm \boldsymbol{b} = (x_1 \boldsymbol{i} + y_1 \boldsymbol{j} + z_1 \boldsymbol{k}) \pm (x_2 \boldsymbol{i} + y_2 \boldsymbol{j} + z_2 \boldsymbol{k})$$
$$= (x_1 \pm x_2) \boldsymbol{i} + (y_1 \pm y_2) \boldsymbol{j} + (z_1 \pm z_2) \boldsymbol{k},$$
$$\lambda \boldsymbol{a} = \lambda (x_1 \boldsymbol{i} + y_1 \boldsymbol{j} + z_1 \boldsymbol{k})$$
$$= (\lambda x_1) \boldsymbol{i} + (\lambda y_1) \boldsymbol{j} + (\lambda z_1) \boldsymbol{k} \quad (\lambda \in \mathbf{R}).$$

所以

$$\boldsymbol{a} \pm \boldsymbol{b} = \{x_1 \pm x_2, y_1 \pm y_2, z_1 \pm z_2\},$$
$$\lambda \boldsymbol{a} = \{\lambda x_1, \lambda y_1, \lambda z_1\}.$$

也就是说,向量的和(差)向量的坐标等于它们的坐标的和(差),数乘向量 $\lambda \boldsymbol{a}$ 的坐标等于数 λ 乘以 $\boldsymbol{a}$ 的坐标.

进一步可得以下结论:

在图 1.10 中,若设空间两点 M, N 的坐标为 $M(x_1, y_1, z_1), N(x_2, y_2, z_2)$,则

(1) **向量坐标**:

$$\overrightarrow{MN} = \overrightarrow{ON} - \overrightarrow{OM}$$
$$= \{x_2, y_2, z_2\} - \{x_1, y_1, z_1\}$$
$$= \{x_2 - x_1, y_2 - y_1, z_2 - z_1\},$$

即向量坐标为终点坐标减去对应的始点坐标.

(2) **向量的模**:

$$|\overrightarrow{MN}| = \sqrt{(x_2 - x_1)^2 + (y_2 - y_1)^2 + (z_2 - z_1)^2}.$$

5) 空间两点间的距离

设 $M(x_1, y_1, z_1), N(x_2, y_2, z_2)$ 为空间两点,则 M 与 N 之间的距离为

$$d = |\overrightarrow{MN}|$$
$$= \sqrt{(x_2 - x_1)^2 + (y_2 - y_1)^2 + (z_2 - z_1)^2}. \qquad (1.2)$$

6) 向量的方向角与方向余弦

非零向量 $\boldsymbol{a}$ 与三条坐标轴正向的夹角 α, β, γ 称为向量 $\boldsymbol{a}$ 的**方向角**.方向角的余弦值称为向量 $\boldsymbol{a}$ 的**方向余弦**.

are called the **direction cosine** of the vector $\boldsymbol{a}$.

If the direction angles of zero vector $\boldsymbol{a}=\{x_1,y_1,z_1\}$ are α,β,γ, then the cosine of the direction angles is

$$\cos\alpha = \frac{x_1}{|\boldsymbol{a}|} = \frac{x_1}{\sqrt{x_1^2+y_1^2+z_1^2}},$$

$$\cos\beta = \frac{y_1}{|\boldsymbol{a}|} = \frac{y_1}{\sqrt{x_1^2+y_1^2+z_1^2}},$$

$$\cos\gamma = \frac{z_1}{|\boldsymbol{a}|} = \frac{z_1}{\sqrt{x_1^2+y_1^2+z_1^2}}.$$

In addition, it is easy to get

$$\cos^2\alpha+\cos^2\beta+\cos^2\gamma=1.$$

Example 1 Proof by vector addition: a quadrilateral with diagonals mutually divided by each other equally is a parallelogram.

Proof Let the diagonal AC,BD of the quadrilateral $ABCD$ intersect at point O, and divide each other equally (Figure 1.12).

As can be seen from the Figure 1.12:
$$\overrightarrow{AB}=\overrightarrow{AO}+\overrightarrow{OB}=\overrightarrow{OC}+\overrightarrow{DO}=\overrightarrow{DC}.$$
Therefore $\overrightarrow{AB}//\overrightarrow{DC}$ and $|\overrightarrow{AB}|=|\overrightarrow{DC}|$, that is quadrilateral $ABCD$ is a parallelogram.

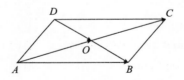

Figure 1.12

Example 2 For vector $\boldsymbol{a}=\{-3,0,1\}$, the coordinates of its starting point A is $(-3,1,4)$, find out the coordinate of its end point B.

Solution Let $B=(x,y,z)$, thus we have
$$\boldsymbol{a}=\{x+3,y-1,z-4\}=\{-3,0,1\},$$
therefore,
$$x=-6,\quad y=1,\quad z=5,$$
that is, the coordinate of B is $(-6,1,5)$.

Example 3 Find a point M on the z-axis whose distances from $A(3,5,-2)$ and $B(4,-1,5)$ are equal.

Solution By the title, we could let the coordinate of M be $(0,0,z)$, and
$$|\overrightarrow{MA}| = |\overrightarrow{MB}|.$$
Derived from formula (1.2), we have
$$\sqrt{3^2+5^2+(-2-z)^2} = \sqrt{(-4)^2+1^2+(5-z)^2}.$$
Thereby $z = \dfrac{2}{7}$, that is, the point is $M\left(0,0,\dfrac{2}{7}\right)$.

Example 4 Let the vectors $\boldsymbol{a} = \{0,-1,2\}, \boldsymbol{b} = \{-1,3,4\}$, find $\boldsymbol{a}+\boldsymbol{b}, 2\boldsymbol{a}-\boldsymbol{b}$.

Solution
$$\begin{aligned}\boldsymbol{a}+\boldsymbol{b} &= \{0+(-1), -1+3, 2+4\} \\ &= \{-1,2,6\},\\ 2\boldsymbol{a}-\boldsymbol{b} &= \{2\times 0, 2\times(-1), 2\times 2\} - \{-1,3,4\} \\ &= \{0-(-1), -2-3, 4-4\} \\ &= \{1,-5,0\}.\end{aligned}$$

1.2 Scalar Product and Cross Product
1.2 数量积与向量积

1. Definition and Operation Law of Scalar Product

Definition 1.10 Let $\boldsymbol{a}, \boldsymbol{b}$ be two vectors, and the quantity $|\boldsymbol{a}||\boldsymbol{b}|\cos\langle \boldsymbol{a},\boldsymbol{b}\rangle$ is called **scalar product** of $\boldsymbol{a}$ and $\boldsymbol{b}$ (also called **inner product** or **dot product**), recorded as $\boldsymbol{a} \cdot \boldsymbol{b}$, read as "$\boldsymbol{a}$ dot product $\boldsymbol{b}$", that is
$$\boldsymbol{a} \cdot \boldsymbol{b} = |\boldsymbol{a}||\boldsymbol{b}|\cos\langle \boldsymbol{a},\boldsymbol{b}\rangle. \tag{1.3}$$

Particularly, when there is a zero vector in the two vectors, there is
$$\boldsymbol{a} \cdot \boldsymbol{b} = 0.$$

Obviously, for coordinate vector $\boldsymbol{i}, \boldsymbol{j}, \boldsymbol{k}$, we have
$$\boldsymbol{i} \cdot \boldsymbol{i} = \boldsymbol{j} \cdot \boldsymbol{j} = \boldsymbol{k} \cdot \boldsymbol{k} = 1,$$
$$\boldsymbol{i} \cdot \boldsymbol{j} = \boldsymbol{j} \cdot \boldsymbol{k} = \boldsymbol{k} \cdot \boldsymbol{i} = 0.$$

By definition, we can obtain:
(1) If $\boldsymbol{b} = \boldsymbol{a}$ in (1.3), there is
$$\boldsymbol{a} \cdot \boldsymbol{a} = |\boldsymbol{a}|^2.$$

We call the scalar product $\boldsymbol{a} \cdot \boldsymbol{a}$ the number squared

$B(-4,1,5)$等距的点M.

解 由题意可设M的坐标为$(0,0,z)$，且
$$|\overrightarrow{MA}| = |\overrightarrow{MB}|.$$
由公式(1.2)得
$$\sqrt{3^2+5^2+(-2-z)^2} = \sqrt{(-4)^2+1^2+(5-z)^2},$$
从而解得$z = \dfrac{2}{7}$，即所求的点为$M\left(0,0,\dfrac{2}{7}\right)$.

例4 设两向量$\boldsymbol{a} = \{0,-1,2\}$和$\boldsymbol{b} = \{-1,3,4\}$，求$\boldsymbol{a}+\boldsymbol{b}, 2\boldsymbol{a}-\boldsymbol{b}$.

解
$$\begin{aligned}\boldsymbol{a}+\boldsymbol{b} &= \{0+(-1), -1+3, 2+4\} \\ &= \{-1,2,6\},\\ 2\boldsymbol{a}-\boldsymbol{b} &= \{2\times 0, 2\times(-1), 2\times 2\} - \{-1,3,4\} \\ &= \{0-(-1), -2-3, 4-4\} \\ &= \{1,-5,0\}.\end{aligned}$$

1. 数量积的定义及运算规律

定义1.10 设$\boldsymbol{a}, \boldsymbol{b}$是两个向量，则数量$|\boldsymbol{a}||\boldsymbol{b}|\cos\langle \boldsymbol{a},\boldsymbol{b}\rangle$称为向量$\boldsymbol{a}$与$\boldsymbol{b}$的**数量积**（也称**内积**或**点积**），记作$\boldsymbol{a} \cdot \boldsymbol{b}$（读作"$\boldsymbol{a}$点乘$\boldsymbol{b}$"），即
$$\boldsymbol{a} \cdot \boldsymbol{b} = |\boldsymbol{a}||\boldsymbol{b}|\cos\langle \boldsymbol{a},\boldsymbol{b}\rangle. \tag{1.3}$$

特别地，当两向量中有一个为零向量时，有
$$\boldsymbol{a} \cdot \boldsymbol{b} = 0.$$

显然，对于坐标向量$\boldsymbol{i}, \boldsymbol{j}, \boldsymbol{k}$，有
$$\boldsymbol{i} \cdot \boldsymbol{i} = \boldsymbol{j} \cdot \boldsymbol{j} = \boldsymbol{k} \cdot \boldsymbol{k} = 1,$$
$$\boldsymbol{i} \cdot \boldsymbol{j} = \boldsymbol{j} \cdot \boldsymbol{k} = \boldsymbol{k} \cdot \boldsymbol{i} = 0.$$

由定义可得到以下结论：
(1) 如果(1.3)式中$\boldsymbol{b} = \boldsymbol{a}$，那么有
$$\boldsymbol{a} \cdot \boldsymbol{a} = |\boldsymbol{a}|^2.$$

我们把数量积$\boldsymbol{a} \cdot \boldsymbol{a}$叫作$\boldsymbol{a}$的数量平方，

of a, and record as a^2, that is
$$a^2 = a \cdot a = |a|^2. \tag{1.4}$$

(2) The two vectors a and b are perpendicular to each other, and the necessary and the sufficient condition is
$$a \cdot b = 0. \tag{1.5}$$

The scalar product of vectors satisfies the following rules of operation: for any vectors a, b, c, and any real numbers λ, μ, there is

(1) **Commutative law**: $a \cdot b = b \cdot a$.

(2) **Distributive law**: $a \cdot (b+c) = a \cdot b + a \cdot c$.

(3) **Combining law** of number factor:
$$(\lambda a) \cdot b = \lambda(a \cdot b) = a \cdot (\lambda b).$$

(4) $a \cdot a = |a|^2 > 0 \; (a \neq 0)$.

(5) $(\lambda a + \mu b) \cdot c = \lambda(a \cdot c) + \mu(b \cdot c)$.

2. Cartesian Coordinate Operation of Scalar Product

Theorem 1.1 In the space Cartesian coordinate system $Oxyz$, let vectors
$$a = x_1 i + y_1 j + z_1 k = \{x_1, y_1, z_1\},$$
$$b = x_2 i + y_2 j + z_2 k = \{x_2, y_2, z_2\},$$
then
$$a \cdot b = x_1 x_2 + y_1 y_2 + z_1 z_2. \tag{1.6}$$

Proof According to the law of the scalar product of vectors, there is
$$\begin{aligned} a \cdot b &= (x_1 i + y_1 j + z_1 k) \cdot (x_2 i + y_2 j + z_2 k) \\ &= x_1 x_2 (i \cdot i) + x_1 y_2 (i \cdot j) + x_1 z_2 (i \cdot k) \\ &\quad + y_1 x_2 (j \cdot i) + y_1 y_2 (j \cdot j) + y_1 z_2 (j \cdot k) \\ &\quad + z_1 x_2 (k \cdot i) + z_1 y_2 (k \cdot j) + z_1 z_2 (k \cdot k) \\ &= x_1 x_2 + y_1 y_2 + z_1 z_2. \end{aligned}$$

Further conclusions are drawn:
Let nonzero vectors $a = \{x_1, y_1, z_1\}$ and $b = \{x_2, y_2, z_2\}$, then

(1) $|a| = \sqrt{a \cdot a}$
$$= \sqrt{x_1^2 + y_1^2 + z_1^2}; \tag{1.7}$$

(2) the cosine of the angle between two vectors a and b is
$$\begin{aligned} \cos\langle a, b \rangle &= \frac{a \cdot b}{|a||b|} \\ &= \frac{x_1 x_2 + y_1 y_2 + z_1 z_2}{\sqrt{x_1^2 + y_1^2 + z_1^2} \cdot \sqrt{x_2^2 + y_2^2 + z_2^2}}. \end{aligned} \tag{1.8}$$

并记为 a^2, 即
$$a^2 = a \cdot a = |a|^2. \tag{1.4}$$

(2) 两向量 a 与 b 相互垂直的充分必要条件是
$$a \cdot b = 0. \tag{1.5}$$

向量的数量积满足下面的运算规律:对于任意向量 a, b, c 及任意实数 λ, μ, 有

(1) **交换律**: $a \cdot b = b \cdot a$.

(2) **分配律**: $a \cdot (b+c) = a \cdot b + a \cdot c$.

(3) **关于数因子的结合律**:
$$(\lambda a) \cdot b = \lambda(a \cdot b) = a \cdot (\lambda b).$$

(4) $a \cdot a = |a|^2 > 0 \; (a \neq 0)$;

(5) $(\lambda a + \mu b) \cdot c = \lambda(a \cdot c) + \mu(b \cdot c)$.

2. 数量积的直角坐标运算

定理1.1 在空间直角坐标系 $Oxyz$ 下,设向量
$$a = x_1 i + y_1 j + z_1 k = \{x_1, y_1, z_1\},$$
$$b = x_2 i + y_2 j + z_2 k = \{x_2, y_2, z_2\},$$
则
$$a \cdot b = x_1 x_2 + y_1 y_2 + z_1 z_2. \tag{1.6}$$

证明 根据向量数量积的运算规律,有
$$\begin{aligned} a \cdot b &= (x_1 i + y_1 j + z_1 k) \cdot (x_2 i + y_2 j + z_2 k) \\ &= x_1 x_2 (i \cdot i) + x_1 y_2 (i \cdot j) + x_1 z_2 (i \cdot k) \\ &\quad + y_1 x_2 (j \cdot i) + y_1 y_2 (j \cdot j) + y_1 z_2 (j \cdot k) \\ &\quad + z_1 x_2 (k \cdot i) + z_1 y_2 (k \cdot j) + z_1 z_2 (k \cdot k) \\ &= x_1 x_2 + y_1 y_2 + z_1 z_2. \end{aligned}$$

进一步可得到以下结论:
设两非零向量 $a = \{x_1, y_1, z_1\}$ 和 $b = \{x_2, y_2, z_2\}$, 则

(1) $|a| = \sqrt{a \cdot a}$
$$= \sqrt{x_1^2 + y_1^2 + z_1^2}; \tag{1.7}$$

(2) 向量 a 与 b 的夹角的余弦为
$$\begin{aligned} \cos\langle a, b \rangle &= \frac{a \cdot b}{|a||b|} \\ &= \frac{x_1 x_2 + y_1 y_2 + z_1 z_2}{\sqrt{x_1^2 + y_1^2 + z_1^2} \cdot \sqrt{x_2^2 + y_2^2 + z_2^2}}. \end{aligned}$$
$$\tag{1.8}$$

3. The Definition and Operation Rule of Cross Product

Definition 1.11 The **cross product** of two vectors ***a*** and ***b*** (also known as the **outer product** or **cross product**) is defined as a vector, recorded as $a \times b$, also as the multiplication cross of ***a*** and ***b***. Its modulus is

$$|a \times b| = |a||b|\sin\theta. \qquad (1.9)$$

Its direction is perpendicular to ***a*** and ***b***, and the right hand system is formed in the order of ***a***, ***b***, $a \times b$.

Obviously, the geometric meaning of the module of the cross product is the area of the parallelogram with the adjacent sides ***a*** and ***b***.

By definition, for the coordinate vector ***i***, ***j***, ***k***, obviously we have

$$i \times i = 0, \quad j \times j = 0, \quad k \times k = 0,$$
$$i \times j = k, \quad j \times k = i, \quad k \times i = j.$$

The operations of cross product satisfy the following rules:

For any vectors ***a***, ***b*** and any real number λ, there is

(1) **Anti commutative law**:
$$a \times b = -b \times a;$$

(2) **Distributive law**:
$$a \times (b+c) = a \times b + a \times c,$$
$$(a+b) \times c = a \times c + b \times c;$$

(3) **The law of union with number multiplication**:
$$(\lambda a) \times b = \lambda(a \times b) = a \times (\lambda b).$$

4. Cartesian Coordinate Operation of Cross Product

Theorem 1.2 In the space Cartesian coordinate system $Oxyz$, let

$$a = x_1 i + y_1 j + z_1 k = \{x_1, y_1, z_1\},$$
$$b = x_2 i + y_2 j + z_2 k = \{x_2, y_2, z_2\},$$

there is

$$a \times b = \begin{vmatrix} y_1 & z_1 \\ y_2 & z_2 \end{vmatrix} i - \begin{vmatrix} x_1 & z_1 \\ x_2 & z_2 \end{vmatrix} j + \begin{vmatrix} x_1 & y_1 \\ x_2 & y_2 \end{vmatrix} k, \qquad (1.10)$$

or

$$a \times b = \begin{vmatrix} i & j & k \\ x_1 & y_1 & z_1 \\ x_2 & y_2 & z_2 \end{vmatrix}. \qquad (1.10)'$$

Proof By the law of operation of the cross product,
$$a \times b = (x_1 i + y_1 j + z_1 k) \times (x_2 i + y_2 j + z_2 k)$$
$$= x_1 x_2 (i \times i) + x_1 y_2 (i \times j) + x_1 z_2 (i \times k)$$
$$+ y_1 x_2 (j \times i) + y_1 y_2 (j \times j) + y_1 z_2 (j \times k)$$
$$+ z_1 x_2 (k \times i) + z_1 y_2 (k \times j) + z_1 z_2 (k \times k)$$
$$= (x_1 y_2 - y_1 x_2)(i \times j) + (y_1 z_2 - z_1 y_2)(j \times k)$$
$$- (x_1 z_2 - z_1 x_2)(k \times i)$$
$$= (y_1 z_2 - z_1 y_2) i - (x_1 z_2 - z_1 x_2) j$$
$$+ (x_1 y_2 - y_1 x_2) k,$$

that is
$$a \times b = \begin{vmatrix} y_1 & z_1 \\ y_2 & z_2 \end{vmatrix} i - \begin{vmatrix} x_1 & z_1 \\ x_2 & z_2 \end{vmatrix} j + \begin{vmatrix} x_1 & y_1 \\ x_2 & y_2 \end{vmatrix} k.$$

The formula above can also be written as
$$a \times b = \begin{vmatrix} i & j & k \\ x_1 & y_1 & z_1 \\ x_2 & y_2 & z_2 \end{vmatrix}.$$

5. The Relationship and Its Judgement of Vectors

According to the definitions of the scalar product and the cross product of the two vectors, the following theorems are easy to obtain.

Theorem 1.3 (Two Vectors Vertical and Its Determination) Let nonzero vectors $a = \{x_1, y_1, z_1\}$, $b = \{x_2, y_2, z_2\}$, there is
$$a \perp b \Leftrightarrow a \cdot b = 0$$
$$\Leftrightarrow x_1 x_2 + y_1 y_2 + z_1 z_2 = 0. \quad (1.11)$$

Theorem 1.4 (Two Vectors Parallelism and Its Determination) Let nonzero vectors $a = \{x_1, y_1, z_1\}$, $b = \{x_2, y_2, z_2\}$, there is
$$a /\!/ b \Leftrightarrow \text{there is a real number } \lambda, \text{ satisfies } a = \lambda b$$
$$\Leftrightarrow a \times b = 0$$
$$\Leftrightarrow \frac{x_1}{x_2} = \frac{y_1}{y_2} = \frac{z_1}{z_2}. \quad (1.12)$$

Rule: In the last formula, the numerator is zero when the denominator is zero.

Theorem 1.5 (Three Vectors Coplanar and Its Determination) Let the vectors $a = \{x_1, y_1, z_1\}$, $b = \{x_2, y_2, z_2\}$, $c = \{x_3, y_3, z_3\}$, there is

a, b, c are coplanar $\Leftrightarrow (a \times b) \cdot c = 0$

$$\Leftrightarrow \begin{vmatrix} x_1 & y_1 & z_1 \\ x_2 & y_2 & z_2 \\ x_3 & y_3 & z_3 \end{vmatrix} = 0. \quad (1.13)$$

Example 1 We have known $|a|=2, |b|=3, \langle a,b \rangle = \frac{2\pi}{3}$, find $a \cdot b, (a-2b) \cdot (a+b), |a+b|$.

Solution According to the scalar product of two vectors, there is

$$a \cdot b = |a||b|\cos\langle a,b \rangle$$
$$= 2 \times 3 \times \cos\frac{2\pi}{3}$$
$$= 2 \times 3 \times \left(-\frac{1}{2}\right) = -3.$$

$$(a-2b) \cdot (a+b) = a \cdot a + a \cdot b - 2b \cdot a - 2b \cdot b$$
$$= |a|^2 - a \cdot b - 2|b|^2$$
$$= 2^2 - (-3) - 2 \times 3^2 = -11.$$

$$|a+b|^2 = (a+b) \cdot (a+b)$$
$$= a \cdot a + a \cdot b + b \cdot a + b \cdot b$$
$$= |a|^2 + 2a \cdot b + |b|^2$$
$$= 2^2 + 2 \times (-3) + 3^2 = 7.$$

So
$$|a+b| = \sqrt{7}.$$

Example 2 Let known points $A(1,-2,3), B(0,1,-2)$, and vector $a = \{4,-1,0\}$, find $a \times \overrightarrow{AB}$ and $\overrightarrow{AB} \times a$.

Solution Because
$$\overrightarrow{AB} = (0-1)i + [1-(-2)]j + (-2-3)k$$
$$= -i + 3j - 5k,$$

that is $\overrightarrow{AB} = \{-1, 3, -5\}$, so

$$a \times \overrightarrow{AB} = \begin{vmatrix} i & j & k \\ 4 & -1 & 0 \\ -1 & 3 & -5 \end{vmatrix}$$
$$= \{5, 20, 11\},$$
$$\overrightarrow{AB} \times a = -a \times \overrightarrow{AB}$$
$$= \{-5, -20, -11\}.$$

Example 3 In the space Cartesian coordinate system $Oxyz$, let $A(4,-1,2), B(1,2,-2), C(2,0,1)$, find the area of $\triangle ABC$.

Solution Apparently, the triangle's area that we are going to find out is half of the area of the parallelogram, which has $\overrightarrow{AB}$ and $\overrightarrow{AC}$ as two of its adjacent sides. According to the geometric meaning of the modulus of cross product of two vectors, we know the area of the parallelogram with $\overrightarrow{AB}$ and $\overrightarrow{AC}$ as its adjacent sides is $|\overrightarrow{AB} \times \overrightarrow{AC}|$, and the area required is $\frac{1}{2}|\overrightarrow{AB} \times \overrightarrow{AC}|$.

Because
$$\overrightarrow{AB} = \{1-4, 2-(-1), -2-2\}$$
$$= \{-3, 3, -4\},$$
$$\overrightarrow{AC} = \{2-4, 0-(-1), 1-2\}$$
$$= \{-2, 1, -1\},$$

we have
$$\overrightarrow{AB} \times \overrightarrow{AC} = \begin{vmatrix} \boldsymbol{i} & \boldsymbol{j} & \boldsymbol{k} \\ -3 & 3 & -4 \\ -2 & 1 & -1 \end{vmatrix}$$
$$= \boldsymbol{i} + 5\boldsymbol{j} + 3\boldsymbol{k}.$$

So
$$|\overrightarrow{AB} \times \overrightarrow{AC}| = \sqrt{1^2 + 5^2 + 3^2} = \sqrt{35}.$$

As a result, the area of △ABC is
$$S_{\triangle ABC} = \frac{1}{2}|\overrightarrow{AB} \times \overrightarrow{AC}| = \frac{\sqrt{35}}{2}.$$

Example 4 Find the unit vector $\boldsymbol{c}$ which is perpendicular to the vector $\boldsymbol{a} = \{2, 2, 1\}$ and $\boldsymbol{b} = \{4, 5, 3\}$ at the same time.

Solution We know that $\boldsymbol{a} \times \boldsymbol{b}$ is perpendicular to $\boldsymbol{a}$ and $\boldsymbol{b}$ at the same time, and $-\boldsymbol{a} \times \boldsymbol{b}$ is perpendicular to $\boldsymbol{a}$ and $\boldsymbol{b}$. We first obtain
$$\boldsymbol{a} \times \boldsymbol{b} = \begin{vmatrix} \boldsymbol{i} & \boldsymbol{j} & \boldsymbol{k} \\ 2 & 2 & 1 \\ 4 & 5 & 3 \end{vmatrix} = \boldsymbol{i} - 2\boldsymbol{j} + 2\boldsymbol{k},$$

then find that
$$|\boldsymbol{a} \times \boldsymbol{b}| = \sqrt{1^2 + (-2)^2 + 2^2} = 3.$$

Therefore there are two unit vectors, that is
$$\boldsymbol{c} = \pm \frac{\boldsymbol{a} \times \boldsymbol{b}}{|\boldsymbol{a} \times \boldsymbol{b}|} = \pm \frac{1}{3}(\boldsymbol{i} - 2\boldsymbol{j} + 2\boldsymbol{k})$$
$$= \pm \left\{\frac{1}{3}, -\frac{2}{3}, \frac{2}{3}\right\}.$$

解 容易看出,所求的面积为以 $\overrightarrow{AB}$, $\overrightarrow{AC}$ 为邻边的平行四边形面积的一半. 而由两向量向量积的模的几何意义知,以 $\overrightarrow{AB}$, $\overrightarrow{AC}$ 为邻边的平行四边形的面积为 $|\overrightarrow{AB} \times \overrightarrow{AC}|$, 从而所求的面积为 $\frac{1}{2}|\overrightarrow{AB} \times \overrightarrow{AC}|$.

由于
$$\overrightarrow{AB} = \{1-4, 2-(-1), -2-2\}$$
$$= \{-3, 3, -4\},$$
$$\overrightarrow{AC} = \{2-4, 0-(-1), 1-2\}$$
$$= \{-2, 1, -1\},$$

因此
$$\overrightarrow{AB} \times \overrightarrow{AC} = \begin{vmatrix} \boldsymbol{i} & \boldsymbol{j} & \boldsymbol{k} \\ -3 & 3 & -4 \\ -2 & 1 & -1 \end{vmatrix}$$
$$= \boldsymbol{i} + 5\boldsymbol{j} + 3\boldsymbol{k},$$

所以
$$|\overrightarrow{AB} \times \overrightarrow{AC}| = \sqrt{1^2 + 5^2 + 3^2} = \sqrt{35}.$$

故△ABC 的面积为
$$S_{\triangle ABC} = \frac{1}{2}|\overrightarrow{AB} \times \overrightarrow{AC}| = \frac{\sqrt{35}}{2}.$$

例 4 求同时垂直于向量 $\boldsymbol{a} = \{2, 2, 1\}$ 和 $\boldsymbol{b} = \{4, 5, 3\}$ 的单位向量 $\boldsymbol{c}$.

解 我们知道, $\boldsymbol{a} \times \boldsymbol{b}$ 同时垂直于 $\boldsymbol{a}$ 和 $\boldsymbol{b}$, 且 $-\boldsymbol{a} \times \boldsymbol{b}$ 也同时垂直于 $\boldsymbol{a}$ 和 $\boldsymbol{b}$. 先求得
$$\boldsymbol{a} \times \boldsymbol{b} = \begin{vmatrix} \boldsymbol{i} & \boldsymbol{j} & \boldsymbol{k} \\ 2 & 2 & 1 \\ 4 & 5 & 3 \end{vmatrix} = \boldsymbol{i} - 2\boldsymbol{j} + 2\boldsymbol{k},$$

再求得
$$|\boldsymbol{a} \times \boldsymbol{b}| = \sqrt{1^2 + (-2)^2 + 2^2} = 3.$$

故所求的单位向量有两个,即
$$\boldsymbol{c} = \pm \frac{\boldsymbol{a} \times \boldsymbol{b}}{|\boldsymbol{a} \times \boldsymbol{b}|} = \pm \frac{1}{3}(\boldsymbol{i} - 2\boldsymbol{j} + 2\boldsymbol{k})$$
$$= \pm \left\{\frac{1}{3}, -\frac{2}{3}, \frac{2}{3}\right\}.$$

1.3 Plane and Its Equation

In this section we use vector theory as a tool to establish the equations of plane in the space Cartesian coordinates. Here are some equations for the plane determined by different conditions.

1. Point Normal form Equation of the Plane

We know that in geometry, a plane that passes a point and is perpendicular to a given direction is one and only one. This geometric relationship can be described by an analytic expression.

As shown in Figure 1.13, if the plane π passes the fixed point $M_0(x_0, y_0, z_0)$ and is perpendicular to the non-zero vector $\boldsymbol{n} = \{A, B, C\}$, π is uniquely identified. Now we are going to find out the equation of the plane π.

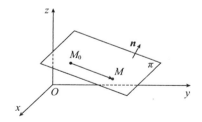

Figure 1.13

Take a point $M(x, y, z)$ which is different from M_0 on the plane π, known by the $\boldsymbol{n} \perp \pi$, so $\boldsymbol{n} \perp \overrightarrow{M_0 M}$. According to sufficient and necessary conditions for the vertical of the two vectors, we can get
$$\boldsymbol{n} \cdot \overrightarrow{M_0 M} = 0.$$
However
$$\overrightarrow{M_0 M} = \{x - x_0, y - y_0, z - z_0\},$$
$$\boldsymbol{n} = \{A, B, C\}.$$

1.3 平面及其方程

本节以向量理论为工具,在空间直角坐标系中建立平面的方程.我们将给出几种由不同条件所确定的平面的方程.

1. 平面的点法式方程

我们知道,在几何上,过空间中某一点且垂直于给定方向的平面有且只有一个.下面用解析式描述此几何关系.

如图 1.13 所示,如果平面 π 过定点 $M_0(x_0, y_0, z_0)$,并且垂直于非零向量 $\boldsymbol{n} = \{A, B, C\}$,则平面 π 被唯一确定.下面来求平面 π 的方程.

图 1.13

任取平面 π 上异于 M_0 的一点 $M(x, y, z)$,由已知有 $\boldsymbol{n} \perp \pi$,因此 $\boldsymbol{n} \perp \overrightarrow{M_0 M}$.由两向量垂直的充分必要条件,可得
$$\boldsymbol{n} \cdot \overrightarrow{M_0 M} = 0.$$
而
$$\overrightarrow{M_0 M} = \{x - x_0, y - y_0, z - z_0\},$$
$$\boldsymbol{n} = \{A, B, C\},$$

So we can get equation
$$A(x-x_0)+B(y-y_0)+C(z-z_0)=0. \quad (1.14)$$

Conversely, if point $M(x,y,z)$ satisfies the equation (1.14), then $n \perp \overrightarrow{M_0M}$. This shows that point M is on plane π. So the equation (1.14) is the equation of plane π, and equation (1.14) is called the **point normal form** of plane π, n is called the **normal vector** of plane π.

Notes Any nonzero n vector which is perpendicular to the plane π can be used as the normal vector of π.

2. General Equation of the Plane

In fact, by the planar point normal form equation (1.14), we can obtain
$$Ax+By+Cz-(Ax_0+By_0+Cz_0)=0. \quad (1.15)$$
Let $D=-(Ax_0+By_0+Cz_0)$, so
$$Ax+By+Cz+D=0 \quad (A,B,C \text{ are all nonzero}),$$
that is, the equations of any planes are the first-order equation of x, y, z. In turn, any one first-order equation of x, y, z represents a plane. So equation (1.15) is a plane equation, which is called the **general equation** of the plane, where $\{A,B,C\}$ is a normal vector of the plane.

Next is several special cases of general equation of the plane:

(1) $D=0$, that is, the equation has a form
$$Ax+By+Cz=0,$$
then the plane passes the origin;

(2) $A=0, D\neq 0$, that is, the equation has a form
$$By+Cz+D=0,$$
then the plane is parallel to the x-axis;

(3) $A=D=0$, that is, the equation has a form
$$By+Cz=0,$$
then the plane passes the x-axis;

(4) $A=B=0, D\neq 0$, that is, the equation has a form
$$Cz+D=0,$$
then the plane is parallel to the Oxy-plane;

(5) $A=B=D=0$, that is, the equation has a form
$$z=0,$$

可得方程
$$A(x-x_0)+B(y-y_0)+C(z-z_0)=0. \quad (1.14)$$

反之，如果点$M(x,y,z)$满足方程(1.14)，那么$n \perp \overrightarrow{M_0M}$。这说明点$M$在平面$\pi$上。所以，方程(1.14)就是平面$\pi$的方程。我们称方程(1.14)为平面$\pi$的**点法式方程**，并称$n=\{A,B,C\}$为平面$\pi$有**法向量**。

注 垂直于平面π的任何非零向量n都可以作为π的法向量。

2. 平面的一般式方程

由平面的点法式方程(1.14)可推出
$$Ax+By+Cz-(Ax_0+By_0+Cz_0)=0, \quad (1.15)$$
设$D=-(Ax_0+By_0+Cz_0)$，则
$$Ax+By+Cz+D=0 \ (A,B,C\text{不全为零}),$$
即任意一个平面的方程都是x,y,z的一次方程。反过来，可以证明任意一个x,y,z的一次方程都表示一个平面。所以说方程(1.15)是一个平面的方程，我们称之为该平面的**一般式方程**，其中$\{A,B,C\}$为该平面的一个法向量。

下面是平面一般方程的几种特殊情况：

（1）$D=0$，即方程具有形式
$$Ax+By+Cz=0,$$
这时平面过原点；

（2）$A=0, D\neq 0$，即方程具有形式
$$By+Cz+D=0,$$
这时平面平行x轴；

（3）$A=D=0$，即方程具有形式
$$By+Cz=0,$$
这时平面过x轴；

（4）$A=B=0, D\neq 0$，即方程具有形式
$$Cz+D=0,$$
这时平面平行于Oxy平面；

（5）$A=B=D=0$，即方程具有形式
$$z=0,$$

then the plane is Oxy-plane.

Similarly, other special situations can be discussed.

3. Intercept Equation of the Plane

Let three points $(a,0,0),(0,b,0),(0,0,c)$ $(abc\neq 0)$ as the intersection of the plane π and three axes (Figure 1.14). Put them into the general equation (1.15) of the plane, we could obtain the equation of the plane π:

$$\frac{x}{a}+\frac{y}{b}+\frac{z}{c}=1. \qquad (1.16)$$

Equation (1.16) is called **intercept equation** of the plane π, and a,b,c are called the **intercepts** of the plane on the x-axis, y-axis and z-axis.

这时平面为 Oxy 平面.

类似地,可讨论其他的特殊情况.

3. 平面的截距式方程

设三点 $(a,0,0),(0,b,0),(0,0,c)$ ($abc\neq 0$) 为平面 π 与三个坐标轴的交点(图 1.14),把它们代入平面的一般式方程 (1.15),可求得平面 π 的方程为

$$\frac{x}{a}+\frac{y}{b}+\frac{z}{c}=1. \qquad (1.16)$$

方程(1.16)称平面 π 的**截距式方程**,其中 a,b,c 分别叫作该平面在 x 轴,y 轴和 z 轴上的**截距**.

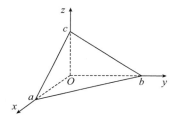

Figure 1.14
图 1.14

4. Three Points Equation of the Plane

Let the plane π pass three points $M_1(x_1,y_1,z_1)$, $M_2(x_2,y_2,z_2)$, $M_3(x_3,y_3,z_3)$, where M_1,M_2,M_3 are not collinear, then it's easy to get the equation of the plane π, that is

$$\begin{vmatrix} x-x_1 & y-y_1 & z-z_1 \\ x_2-x_1 & y_2-y_1 & z_2-z_1 \\ x_3-x_1 & y_3-y_1 & z_3-z_1 \end{vmatrix}=0. \qquad (1.17)$$

Equation (1.17) is called **three points equation** of the plane π.

In fact, the plane π passes three non collinear points M_1,M_2,M_3, and its normal vector is $\overrightarrow{M_1M_2}\times\overrightarrow{M_1M_3}\neq \mathbf{0}$. Take a point $M(x,y,z)$ on the plane π, then we can get the plane equation

$$\overrightarrow{M_1M}\cdot(\overrightarrow{M_1M_2}\times\overrightarrow{M_1M_3})=0.$$

Thus the three points equation of the plane is (1.17).

4. 平面的三点式方程

设平面 π 过三点 $M_1(x_1,y_1,z_1)$, $M_2(x_2,y_2,z_2)$, $M_3(x_3,y_3,z_3)$,且 M_1,M_2,M_3 不共线,容易导出平面 π 的方程为

$$\begin{vmatrix} x-x_1 & y-y_1 & z-z_1 \\ x_2-x_1 & y_2-y_1 & z_2-z_1 \\ x_3-x_1 & y_3-y_1 & z_3-z_1 \end{vmatrix}=0. \qquad (1.17)$$

称方程(1.17)为平面 π 的**三点式方程**.

事实上,平面 π 过不共线三点 M_1,M_2,M_3,可取法向量为 $\overrightarrow{M_1M_2}\times\overrightarrow{M_1M_3}\neq \mathbf{0}$,再取平面 π 上任一点 $M(x,y,z)$,则得平面方程为

$$\overrightarrow{M_1M}\cdot(\overrightarrow{M_1M_2}\times\overrightarrow{M_1M_3})=0.$$

代入向量的坐标,化简可得平面的三点式方程(1.17).

5. The Angle Between Two Planes and the Positional Relationship

Definition 1.12 Define the **angle** θ between two planes as the normal vector of two planes (Figure 1.15), and define $0 \leqslant \theta \leqslant \dfrac{\pi}{2}$.

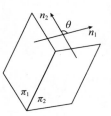

Figure 1.15

Let the equations of two planes π_1 and π_2 are
$$\pi_1: A_1 x + B_1 y + C_1 z + D_1 = 0,$$
$$\pi_2: A_2 x + B_2 y + C_2 z + D_2 = 0,$$
where A_1, B_1, C_1 are all nonzero and A_2, B_2, C_2 are all nonzero, then their normal vectors are $\boldsymbol{n}_1 = \{A_1, B_1, C_1\}$ and $\boldsymbol{n}_2 = \{A_2, B_2, C_2\}$. Thus their angle is
$$\theta = \langle \boldsymbol{n}_1, \boldsymbol{n}_2 \rangle \quad \text{or} \quad \theta = \pi - \langle \boldsymbol{n}_1, \boldsymbol{n}_2 \rangle.$$
So, according to the cosine formula of the angle between two vectors, we get
$$\cos\theta = \cos\langle \boldsymbol{n}_1, \boldsymbol{n}_2 \rangle = \frac{|\boldsymbol{n}_1 \cdot \boldsymbol{n}_2|}{|\boldsymbol{n}_1||\boldsymbol{n}_2|}$$
$$= \frac{|A_1 A_2 + B_1 B_2 + C_1 C_2|}{\sqrt{A_1^2 + B_1^2 + C_1^2} \cdot \sqrt{A_2^2 + B_2^2 + C_2^2}}. \quad (1.18)$$

We can derive the positional relationship from the relationship between the normal vectors of two plane equations:

(1) π_1 and π_2 are coincident $\Leftrightarrow \dfrac{A_1}{A_2} = \dfrac{B_1}{B_2} = \dfrac{C_1}{C_2} = \dfrac{D_1}{D_2}$;

(2) $\pi_1 \parallel \pi_2 \Leftrightarrow \boldsymbol{n}_1 \parallel \boldsymbol{n}_2$
$\Leftrightarrow \dfrac{A_1}{A_2} = \dfrac{B_1}{B_2} = \dfrac{C_1}{C_2} \neq \dfrac{D_1}{D_2}$;

(3) π_1 and π_2 are intersected $\Leftrightarrow A_1, B_1, C_1$ is out of proportion to A_2, B_2, C_2.

5. 两平面的夹角和位置关系

定义 1.12 定义两平面的**夹角** θ 为两平面的法向量的夹角，并规定 $0 \leqslant \theta \leqslant \dfrac{\pi}{2}$ (图 1.15).

设两个平面 π_1 与 π_2 的方程分别为
$$\pi_1: A_1 x + B_1 y + C_1 z + D_1 = 0,$$
$$\pi_2: A_2 x + B_2 y + C_2 z + D_2 = 0,$$
其中 A_1, B_1, C_1 不同时为零，且 A_2, B_2, C_2 不同时为零，则它们的法向量分别为 $\boldsymbol{n}_1 = \{A_1, B_1, C_1\}$ 和 $\boldsymbol{n}_2 = \{A_2, B_2, C_2\}$，从而它们的夹角为
$$\theta = \langle \boldsymbol{n}_1, \boldsymbol{n}_2 \rangle \quad \text{或} \quad \theta = \pi - \langle \boldsymbol{n}_1, \boldsymbol{n}_2 \rangle.$$
于是，根据两向量夹角的余弦公式，得
$$\cos\theta = \cos\langle \boldsymbol{n}_1, \boldsymbol{n}_2 \rangle = \frac{|\boldsymbol{n}_1 \cdot \boldsymbol{n}_2|}{|\boldsymbol{n}_1||\boldsymbol{n}_2|}$$
$$= \frac{|A_1 A_2 + B_1 B_2 + C_1 C_2|}{\sqrt{A_1^2 + B_1^2 + C_1^2} \cdot \sqrt{A_2^2 + B_2^2 + C_2^2}}.$$
$$(1.18)$$

进一步，我们可以从两个平面方程的法向量之间的关系导出它们之间的位置关系：

(1) π_1 与 π_2 重合 $\Leftrightarrow \dfrac{A_1}{A_2} = \dfrac{B_1}{B_2} = \dfrac{C_1}{C_2} = \dfrac{D_1}{D_2}$;

(2) $\pi_1 \parallel \pi_2 \Leftrightarrow \boldsymbol{n}_1 \parallel \boldsymbol{n}_2$
$\Leftrightarrow \dfrac{A_1}{A_2} = \dfrac{B_1}{B_2} = \dfrac{C_1}{C_2} \neq \dfrac{D_1}{D_2}$;

(3) π_1 与 π_2 相交 $\Leftrightarrow A_1, B_1, C_1$ 与 A_2, B_2, C_2 不成比例.

(4) $\pi_1 \perp \pi_2 \Leftrightarrow \boldsymbol{n}_1 \perp \boldsymbol{n}_2 \Leftrightarrow \boldsymbol{n}_1 \cdot \boldsymbol{n}_2 = 0$
$\Leftrightarrow A_1A_2 + B_1B_2 + C_1C_2 = 0.$ (1.19)

6. Distance from Point to Plane

In the space Cartesian coordinate system $Oxyz$, let the plane $\pi: Ax + By + Cz + D = 0$ (A, B, C are all nonzero), and point $M_0(x_0, y_0, z_0)$ is outside the plane, then the distance from point M_0 to the plane is

$$d = \frac{|Ax_0 + By_0 + Cz_0 + D|}{\sqrt{A^2 + B^2 + C^2}}. \quad (1.20)$$

Example 1 Find a plane equation that passes the point $M_0(1, 3, 5)$ and is parallel to the Oxy-plane.

Solution It is obvious that $\boldsymbol{n} = \{0, 0, 1\}$ is a normal vector of the plane, so the equation of the plane is
$$0 \cdot (x - 1) + 0 \cdot (y - 3) + 1 \cdot (z - 5) = 0,$$
that is $z - 5 = 0.$

Example 2 Find a plane equation that passes the two points $(7, 5, -2)$ and $(-1, 4, 3)$, and the intercept on x-axis is 5.

Solution The plane passes three points $(7, 5, -2)$, $(-1, 4, 3)$, $(5, 0, 0)$, by the three points equation, the plane equation is
$$\begin{vmatrix} x-5 & y & z \\ 2 & 5 & -2 \\ -6 & 4 & 3 \end{vmatrix} = 0,$$
that is $23x + 6y + 38z - 115 = 0.$

Example 3 Find a plane that passes the two points $A(2, 0, 1)$, $B(9, 6, 1)$, and is parallel to the z-axis.

Solution The normal vector of the plane is
$$\boldsymbol{n} = \overrightarrow{AB} \times \boldsymbol{k} = \{7, 6, 0\} \times \{0, 0, 1\}$$
$$= \{6, -7, 0\},$$
and because the plane passes the point $A(2, 0, 1)$, the equation of the required plane is
$$6(x - 2) - 7(y - 0) = 0,$$
that is $6x - 7y - 12 = 0.$

Example 4 Find a plane equation that passes the point $(7, -5, 1)$ and x-axis.

Solution The plane passes x-axis, so $A = D = 0$. Thus

(4) $\pi_1 \perp \pi_2 \Leftrightarrow \boldsymbol{n}_1 \perp \boldsymbol{n}_2 \Leftrightarrow \boldsymbol{n}_1 \cdot \boldsymbol{n}_2 = 0$
$\Leftrightarrow A_1A_2 + B_1B_2 + C_1C_2 = 0.$
(1.19)

6. 点到平面的距离

在空间直角坐标系 $Oxyz$ 中,设有平面 $\pi: Ax + By + Cz + D = 0$($A, B, C$ 不全为零)及平面外的一点 $M_0(x_0, y_0, z_0)$,则可以推导出点 M_0 到平面的距离为

$$d = \frac{|Ax_0 + By_0 + Cz_0 + D|}{\sqrt{A^2 + B^2 + C^2}}. \quad (1.20)$$

例1 求通过点 $M_0(1, 3, 5)$ 且与 Oxy 平面平行的平面方程.

解 显然 $\boldsymbol{n} = \{0, 0, 1\}$ 为所求的平面一个法向量,因此所求的平面方程为
$$0 \cdot (x - 1) + 0 \cdot (y - 3) + 1 \cdot (z - 5) = 0,$$
即 $z - 5 = 0.$

例2 求过两个定点 $(7, 5, -2)$ 和 $(-1, 4, 3)$ 且在 x 轴上的截距是 5 的平面方程.

解 所求的平面过三点 $(7, 5, -2)$,$(-1, 4, 3)$,$(5, 0, 0)$,由三点式,所求的平面方程为
$$\begin{vmatrix} x-5 & y & z \\ 2 & 5 & -2 \\ -6 & 4 & 3 \end{vmatrix} = 0,$$
即 $23x + 6y + 38z - 115 = 0.$

例3 求过两个定点 $A(2, 0, 1)$,$B(9, 6, 1)$ 且平行于 z 轴的平面.

解 所求平面的法向量为
$$\boldsymbol{n} = \overrightarrow{AB} \times \boldsymbol{k} = \{7, 6, 0\} \times \{0, 0, 1\}$$
$$= \{6, -7, 0\}.$$
又该平面过点 $A(2, 0, 1)$,故所求平面的方程为
$$6(x - 2) - 7(y - 0) = 0,$$
即 $6x - 7y - 12 = 0.$

例4 求过定点 $(7, -5, 1)$ 且过 x 轴的平面方程.

解 所求的平面过 x 轴,故 $A = D = 0$,

we let the plane equation is
$$By+Cz=0.$$
The plane also passes the point $(7,-5,1)$, so we have
$$-5B+C=0.$$
Let $B=1$, we obtain $C=5$. Therefore the plane equation is
$$y+5z=0.$$

Example 5 Find the distance between two parallel planes $x-2y+3z+1=0$ and $x-2y+3z-2=0$.

Solution Take one point on plane $x-2y+3z+1=0$, such as $M(-1,0,0)$. Then the distance from M to another plane is the distance between two parallel planes. So the distance between two parallel planes is
$$d=\frac{|1\times(-1)-2|}{\sqrt{1^2+(-2)^2+3^2}}=\frac{3}{\sqrt{14}}=\frac{3\sqrt{14}}{14}.$$

1.4 Space Straight Lines and Their Equations
1.4 空间直线及其方程

This section mainly describes how to establish several different forms of linear equations in the Cartesian coordinate system.

本节主要介绍如何建立空间直角坐标系中几种不同形式的直线方程.

1. Symmetric Equation of a Straight Line

We know that in geometry we can uniquely determine a straight line with one known point and in one direction in a given space. The following analysis is used to describe this geometry.

As shown in Figure 1.16, in the space Cartesian coordinate system $Oxyz$, if straight line passes the point $M_0(x_0,y_0,z_0)$ and is parallel to non-zero vector $\boldsymbol{v}=\{X,Y,Z\}$, then the straight line l is uniquely determined. Let's create the equation of the straight line.

Let $M(x,y,z)$ for any point other than the point M_0 on a line l, therefore $\overrightarrow{M_0M}\parallel\boldsymbol{v}$. The coordinates of
$$\overrightarrow{M_0M}=\{x-x_0,y-y_0,z-z_0\},$$

1. 直线的对称式方程

我们知道,在几何上,给定空间中某一点和一个方向可以唯一确定一条直线.下面用解析式描述此几何关系.

如图 1.16 所示,在空间直角坐标系 $Oxyz$ 中,如果直线过点 $M_0(x_0,y_0,z_0)$,并与非零向量 $\boldsymbol{v}=\{X,Y,Z\}$ 平行,那么直线 l 被唯一确定.下面我们来建立直线 l 的方程.

设 $M(x,y,z)$ 为直线 l 上异于 M_0 的任一点,则 $\overrightarrow{M_0M}\parallel\boldsymbol{v}$.由于
$$\overrightarrow{M_0M}=\{x-x_0,y-y_0,z-z_0\},$$

1.4 Space Straight Lines and Their Equations
1.4 直线及其方程

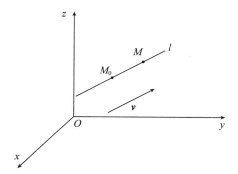

Figure 1.16
图 1.16

therefore by the necessary and sufficient conditions for two parallel vector, there is

$$\frac{x-x_0}{X}=\frac{y-y_0}{Y}=\frac{z-z_0}{Z}. \quad (1.21)$$

Conversely, if the coordinates of point $M(x,y,z)$ satisfy equation (1.21), then $\overrightarrow{M_0M}$ is parallel to v, so the point M is on line l. Therefore, equation (1.21) is the equation of l, which is usually called the **symmetry equation** of l (or **point oriented equation**), and the vector v is called the **direction vector** of line l.

Remark (1) Any nonzero vectors parallel to the line l can be used as a direction vector of the line l.

(2) Because of the direction vector $v \neq 0$, that is X,Y,Z are all nonzero, we have the following special cases:

(i) When one of X,Y,Z is a zero, such as $X=0$, equation (1.21) can be translated into

$$\begin{cases} x-x_0=0, \\ \dfrac{y-y_0}{Y}=\dfrac{z-z_0}{Z}. \end{cases}$$

That is, the line is parallel to the Oyz-plane.

(ii) When two of X,Y,Z are zero, such as $X=Y=0$, equation (1.21) can be translated into

$$\begin{cases} x-x_0=0, \\ y-y_0=0, \end{cases}$$

that is, the line is parallel to the z-axis.

因此由两向量平行的充分必要条件有

$$\frac{x-x_0}{X}=\frac{y-y_0}{Y}=\frac{z-z_0}{Z}. \quad (1.21)$$

反之,如果点 $M(x,y,z)$ 的坐标满足方程(1.21),那么说明 $\overrightarrow{M_0M}$ 与 v 平行,于是点 M 在直线 l 上. 所以,方程(1.21)就是直线 l 的方程,通常称之为直线 l 的**对称式方程**(或**点向式方程**),其中向量 v 称为直线 l 的**方向向量**.

注 (1) 与直线 l 平行的任一非零向量都可作为直线 l 的方向向量.

(2) 由于直线 l 的方向向量 $v \neq \mathbf{0}$,即 X,Y,Z 不全为零,所以有以下特殊情形:

(i) 当 X,Y,Z 中有一个为零时,如 $X=0$ 时,(1.21)式转化为

$$\begin{cases} x-x_0=0, \\ \dfrac{y-y_0}{Y}=\dfrac{z-z_0}{Z}, \end{cases}$$

即该直线与 Oyz 平面平行;

(ii) 当 X,Y,Z 中有两个为零时,如 $X=Y=0$ 时,(1.21)式转化为

$$\begin{cases} x-x_0=0, \\ y-y_0=0, \end{cases}$$

即该直线与 z 轴平行.

2. Parametric Equation of a Straight Line

In the space Cartesian coordinate system $Oxyz$, let a point $M_0(x_0, y_0, z_0)$ on the line l, and $\boldsymbol{v} = \{X, Y, Z\}$ is a direction vector of l, then the straight line l has a symmetric equation of (1.21). If the symmetric equation of a straight line is

$$\frac{x-x_0}{X} = \frac{y-y_0}{Y} = \frac{z-z_0}{Z} = t,$$

then we have

$$\begin{cases} x = x_0 + Xt, \\ y = y_0 + Yt, \\ z = z_0 + Zt. \end{cases} \qquad (1.22)$$

Equation (1.22) is called **parametric equation** of a straight line.

3. General Equation of a Straight Line

The equations of two non-parallel planes π_1, π_2 are

$$\pi_1: A_1 x + B_1 y + C_1 z + D_1 = 0,$$
$$\pi_2: A_2 x + B_2 y + C_2 z + D_2 = 0,$$

where A_1, B_1, C_1 is out of proportion to A_2, B_2, C_2. So intersection of these two planes must be a straight line, written l as

$$l: \begin{cases} A_1 x + B_1 y + C_1 z + D_1 = 0, \\ A_2 x + B_2 y + C_2 z + D_2 = 0. \end{cases} \qquad (1.23)$$

Obviously, the coordinates at any point on the line l satisfy equation (1.23). Conversely, if the coordinates of $M(x, y, z)$ satisfy equation (1.23), then point M is on planes π_1 and π_2 at the same time, and thus it is also on the intersection l of π_1 and π_2. So equation (1.23) is the equation of line l, which is called the **general equation** of line l.

4. The General Formula of Linear Equation and Transformation of Symmetric Formula

We know, a general equation of a known linear equation, such as (1.23). We could transforme it into a symmetric equation.

In fact, the normal vectors of the two planes in the formula (1.23) are $\boldsymbol{n}_1 = \{A_1, B_1, C_1\}$, $\boldsymbol{n}_2 = \{A_2, B_2, C_2\}$.

Because $\boldsymbol{n}_1 \perp l, \boldsymbol{n}_2 \perp l \Rightarrow \boldsymbol{n}_1 \times \boldsymbol{n}_2$, we can take

$$\boldsymbol{n}_1 \times \boldsymbol{n}_2 = \left\{ \begin{vmatrix} B_1 & C_1 \\ B_2 & C_2 \end{vmatrix}, \begin{vmatrix} C_1 & A_1 \\ C_2 & A_2 \end{vmatrix}, \begin{vmatrix} A_1 & B_1 \\ A_2 & B_2 \end{vmatrix} \right\}$$

as the direction vector of line l. Take a point in the line and let it to $M_0(x_0, y_0, z_0)$, so the symmetric equation of the straight line is

$$\frac{x-x_0}{\begin{vmatrix} B_1 & C_1 \\ B_2 & C_2 \end{vmatrix}} = \frac{y-y_0}{\begin{vmatrix} C_1 & A_1 \\ C_2 & A_2 \end{vmatrix}} = \frac{z-z_0}{\begin{vmatrix} A_1 & B_1 \\ A_2 & B_2 \end{vmatrix}}.$$

A symmetric formula of line l, such as equation (1.21), could be transformed into a general equation.

For example, let's suppose none of the coordinate components X, Y and Z of the direction vector in equation (1.21) are equal to zero. Hence we can obtain two sub-equations from the equation above, that is

$$\begin{cases} \dfrac{x-x_0}{X} = \dfrac{y-y_0}{Y}, \\ \dfrac{y-y_0}{Y} = \dfrac{z-z_0}{Z}. \end{cases}$$

We could obtain the general equation of straight line l shaped as (1.23).

5. The Angle and Positional Relation Between Two Straight Lines in Space

Definition 1.13 Define the **angle** θ between two straight lines as the direction vector of the straight line, and fix that $0 \leqslant \theta \leqslant \dfrac{\pi}{2}$.

The equations for two straight lines are

$$l_1: \frac{x-x_1}{X_1} = \frac{y-y_1}{Y_1} = \frac{z-z_1}{Z_1},$$

$$l_2: \frac{x-x_2}{X_2} = \frac{y-y_2}{Y_2} = \frac{z-z_2}{Z_2}.$$

We could see that a direction vector of line l_1 is $\boldsymbol{v}_1 = \{X_1, Y_1, Z_1\}$, point $M_1(x_1, y_1, z_1)$ is the point of line l_1; a direction vector of line l_2 is $\boldsymbol{v}_2 = \{X_2, Y_2, Z_2\}$, point $M_2(x_2, y_2, z_2)$ is the point of line l_2. Then the angle between the line l_1 and l_2 is $\theta = \langle \boldsymbol{v}_1, \boldsymbol{v}_2 \rangle$ or $\theta = \pi - \langle \boldsymbol{v}_1, \boldsymbol{v}_2 \rangle$, and further we have

---

为 $\boldsymbol{n}_1 \perp l, \boldsymbol{n}_2 \perp l \Rightarrow \boldsymbol{n}_1 \times \boldsymbol{n}_2$, 所以可取

$$\boldsymbol{n}_1 \times \boldsymbol{n}_2 = \left\{ \begin{vmatrix} B_1 & C_1 \\ B_2 & C_2 \end{vmatrix}, \begin{vmatrix} C_1 & A_1 \\ C_2 & A_2 \end{vmatrix}, \begin{vmatrix} A_1 & B_1 \\ A_2 & B_2 \end{vmatrix} \right\}$$

为直线 l 的方向向量. 在直线 l 上任取一点 $M_0(x_0, y_0, z_0)$, 于是直线 l 的对称式方程为

$$\frac{x-x_0}{\begin{vmatrix} B_1 & C_1 \\ B_2 & C_2 \end{vmatrix}} = \frac{y-y_0}{\begin{vmatrix} C_1 & A_1 \\ C_2 & A_2 \end{vmatrix}} = \frac{z-z_0}{\begin{vmatrix} A_1 & B_1 \\ A_2 & B_2 \end{vmatrix}}.$$

若已知直线 l 的对称式方程为(1.21), 我们同样也可将其转化成一般式方程.

例如, 设方程(1.21)中方向向量的坐标分量 X, Y, Z 都不等于 0, 将方程分为两个等式, 得

$$\begin{cases} \dfrac{x-x_0}{X} = \dfrac{y-y_0}{Y}, \\ \dfrac{y-y_0}{Y} = \dfrac{z-z_0}{Z}. \end{cases}$$

整理即可得到形如(1.23)式的直线 l 的一般式方程.

5. 空间中两直线的夹角和位置关系

定义 1.13 定义两直线的**夹角** θ 为两直线的方向向量的夹角, 且规定 $0 \leqslant \theta \leqslant \dfrac{\pi}{2}$.

设两直线 l_1 与 l_2 的方程分别为

$$l_1: \frac{x-x_1}{X_1} = \frac{y-y_1}{Y_1} = \frac{z-z_1}{Z_1},$$

$$l_2: \frac{x-x_2}{X_2} = \frac{y-y_2}{Y_2} = \frac{z-z_2}{Z_2}.$$

可见, 直线 l_1 的一个方向向量为 $\boldsymbol{v}_1 = \{X_1, Y_1, Z_1\}$, 点 $M_1(x_1, y_1, z_1)$ 为直线 l_1 上的点; 直线 l_2 的一个方向向量为 $\boldsymbol{v}_2 = \{X_2, Y_2, Z_2\}$, 点 $M_2(x_2, y_2, z_2)$ 为直线 l_2 上的点. 于是, 直线 l_1 与 l_2 的夹角为 $\theta = \langle \boldsymbol{v}_1, \boldsymbol{v}_2 \rangle$ 或 $\theta = \pi - \langle \boldsymbol{v}_1, \boldsymbol{v}_2 \rangle$, 进而可得

$$\cos\theta = |\cos\langle \boldsymbol{v}_1, \boldsymbol{v}_2\rangle| = \frac{|\boldsymbol{v}_1 \cdot \boldsymbol{v}_2|}{|\boldsymbol{v}_1||\boldsymbol{v}_2|}$$
$$= \frac{|x_1 x_2 + y_1 y_2 + z_1 z_2|}{\sqrt{x_1^2 + y_1^2 + z_1^2}\sqrt{x_2^2 + y_2^2 + z_2^2}}. \quad (1.24)$$

The l_1 and l_2 have the following positional relationships:

(1) $l_1 /\!/ l_2 \Leftrightarrow \boldsymbol{v}_1 /\!/ \boldsymbol{v}_2 \Leftrightarrow \dfrac{X_1}{X_2} = \dfrac{Y_1}{Y_2} = \dfrac{Z_1}{Z_2}$

(Here we consider that superposition is the special case of parallel lines);

(2) $l_1 \perp l_2 \Leftrightarrow \boldsymbol{v}_1 \perp \boldsymbol{v}_2$
$\Leftrightarrow X_1 X_2 + Y_1 Y_2 + Z_1 Z_2 = 0.$

6. The Angle and Position Relation Between a Line and a Plane

In the space, there are three types of the position relation between a line and a plane: the line is in the plane, the line is parallel to the plane, the line intersects the plane. Their positional relationship can be determined by the relation between the normal vector of the plane and the direction vector of the line.

Definition 1.14 Suppose the straight line l is not perpendicular to the plane π, and the **angle** between the straight line l and the plane π is defined as the angle between the straight line l and its projected straight line on the plane π. When the straight line l is perpendicular to plane π, the provisions of the angle between π and l is $\dfrac{\pi}{2}$ (Figure 1.17).

进一步，关于直线 l_1 与 l_2 的位置关系，有如下结论：

(1) $l_1 /\!/ l_2 \Leftrightarrow \boldsymbol{v}_1 /\!/ \boldsymbol{v}_2 \Leftrightarrow \dfrac{X_1}{X_2} = \dfrac{Y_1}{Y_2} = \dfrac{Z_1}{Z_2}$

（这里将重合看作平行的特殊情形）；

(2) $l_1 \perp l_2 \Leftrightarrow \boldsymbol{v}_1 \perp \boldsymbol{v}_2$
$\Leftrightarrow X_1 X_2 + Y_1 Y_2 + Z_1 Z_2 = 0.$

6. 直线与平面的夹角和位置关系

在空间中，直线与平面的位置关系有三种：直线在平面上，直线与平面平行，直线与平面相交. 直线与平面位置关系可以通过平面的法向量与直线的方向向量的关系来判定.

定义 1.14 设直线 l 与平面 π 不垂直，定义直线 l 与平面 π 的**夹角**为直线 l 与它在平面 π 上的投影直线的夹角. 当直线 l 与平面 π 垂直时，l 与 π 的夹角为 $\dfrac{\pi}{2}$（图 1.17）.

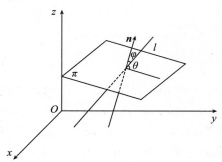

Figure 1.17
图 1.17

The equation of straight line l is
$$\frac{x-x_0}{X}=\frac{y-y_0}{Y}=\frac{z-z_0}{Z},$$
that is l crosses point $M_0(x_0, y_0, z_0)$, and its direction vector is $\boldsymbol{v}=\{X, Y, Z\}$. Also let the equation of plane π is
$$Ax+By+Cz+D=0,$$
that is normal vector of plane π is $\boldsymbol{n}=\{A, B, C\}$, as shown in Figure 1.17, if l is not perpendicular to π, then $\theta = \left|\frac{\pi}{2}-\langle\boldsymbol{v}, \boldsymbol{n}\rangle\right|$. So we have

$$\sin\theta = |\cos\langle\boldsymbol{v}, \boldsymbol{n}\rangle| = \frac{|\boldsymbol{v}\cdot\boldsymbol{n}|}{|\boldsymbol{v}|\cdot|\boldsymbol{n}|}$$
$$= \frac{|AX+BY+CZ|}{\sqrt{X^2+Y^2+Z^2}\sqrt{A^2+B^2+C^2}}. \quad (1.25)$$

Further more, for the linear relationship between position l and plane π, we have the following conclusions:

(1) $l/\!/\pi \Leftrightarrow \boldsymbol{v}\perp\boldsymbol{n} \Leftrightarrow XA+YB+ZC=0$
(Here line l in plane π is a special case of $l/\!/\pi$);

(2) $l\perp\pi \Leftrightarrow \boldsymbol{v}/\!/\boldsymbol{n} \Leftrightarrow \boldsymbol{v}\times\boldsymbol{n}=\boldsymbol{0}$
$\Leftrightarrow \dfrac{X}{A}=\dfrac{Y}{B}=\dfrac{Z}{C}$.

7. Distance from Point to Line

In the space Cartesian coordinate system $Oxyz$, let straight line
$$l: \frac{x-x_1}{X}=\frac{y-y_1}{Y}=\frac{z-z_1}{Z},$$
that is a direction vector of line l is $\boldsymbol{v}=\{X, Y, Z\}$, and l passes point $M_1=\{x_1, y_1, z_1\}$. Let point $M_0(x_0, y_0, z_0)$ outside the line. We could obtain the distance from the point M_0 to the line is
$$d = \frac{|\boldsymbol{v}\times\overrightarrow{M_1M_0}|}{|\boldsymbol{v}|}$$
$$= \frac{\sqrt{\left|\begin{array}{cc}y_0-y_1 & z_0-z_1 \\ Y & Z\end{array}\right|^2 + \left|\begin{array}{cc}z_0-z_1 & x_0-x_1 \\ Z & X\end{array}\right|^2 + \left|\begin{array}{cc}x_0-x_1 & y_0-y_1 \\ X & Y\end{array}\right|^2}}{\sqrt{X^2+Y^2+Z^2}}. \quad (1.26)$$

Example 1 Find a straight line equation of two points $(1, 0, -2)$ and $(0, 2, 3)$.

Solution The direction vector of the straight line is
$$v = \{1-0, 0-2, -2-3\}$$
$$= \{1, -2, -5\},$$
so the equation of straight line is
$$\frac{x-1}{1} = \frac{y}{-2} = \frac{z+2}{-5}.$$

Example 2 Find an equation of straight line in which the point $M(0,1,-3)$ is found, and which is parallel to the two planes
$$\pi_1: y+z=1 \quad \text{and} \quad \pi_2: x-2y+4z=5.$$

Solution The desired line is parallel to planes π_1 and π_2, that is, perpendicular to the normal vectors n_1, n_2 of π_1, π_2, among
$$n_1 = \{0,1,1\}, \quad n_2 = \{1,-2,4\}.$$
Therefore, $n_1 \times n_2$ can be used as a direction vector of the straight line v, then
$$v = n_1 \times n_2 = \begin{vmatrix} i & j & k \\ 0 & 1 & 1 \\ 1 & -2 & 4 \end{vmatrix} = 6i+j-k,$$
that is $v = \{6,1,-1\}$. Thus the linear equation is
$$\frac{x}{6} = \frac{y-1}{1} = \frac{z+3}{-1}.$$

Example 3 Let the linear equation $\begin{cases} x+y+z=5, \\ 3x-3y+5z=7 \end{cases}$ translate into symmetric equation.

Solution A general formula for linear equations can be obtained
$$n_1 \times n_2 = \{1,1,1\} \times \{3,-3,5\}$$
$$= 2\{4,-1,-3\},$$
where n_1, n_2 are the normal vectors of the plane represented by the first and second equations, respectively. Thus, the direction vector of the line is taken as $v = \{4,-1,-3\}$. The equation of given straight line can be used to find a fixed point $\left(0, \frac{9}{4}, \frac{11}{4}\right)$ on the straight line, so the symmetric equation of the straight line is
$$\frac{x}{4} = \frac{y-\frac{9}{4}}{-1} = \frac{z-\frac{11}{4}}{-3}.$$

1.5 Quadratic Surfaces and Their Equations

Similar to the definition of curves and equations in plane analytic geometry, the equations of space surfaces can be defined.

Definition 1.15 If the surface Σ and the equation
$$F(x,y,z)=0$$
satisfy:

(1) The coordinates of each point on the surface Σ satisfies the equation $F(x,y,z)=0$;

(2) The points whose coordinates are the solution of $F(x,y,z)=0$ are all on the surface Σ.

The equation $F(x,y,z)=0$ is called the **equation of surface** Σ, and the surface Σ is called the **graph of this equation.**

In particular, if the equation $F(x,y,z)$ of the surface Σ is a quadratic equation with respect to x,y,z, the surface Σ is called a **quadratic surface.**

This section mainly introduces several common quadratic surface, including spherical surface, ellipsoid, hyperboloid, paraboloid, some cylindrical surfaces and rotate surface.

1. Spherical Surface

In space, the distance from a given point is equal to the trajectory of a fix-length point, and its trajectory is **spherical surface.** The fixed point is called the **center of the sphere**, and the fixed length is called the **radius** of the sphere.

We can easily deduce that point $C(x_0,y_0,z_0)$ is the center, the radius is fixed length R, its spherical surface is
$$(x-x_0)^2+(y-y_0)^2+(z-z_0)^2=R^2. \quad (1.27)$$

In fact, if $M(x,y,z)$ is the point of the spherical surface, there is $|\overrightarrow{MC}|=R$, that is
$$\sqrt{(x-x_0)^2+(y-y_0)^2+(z-z_0)^2}=R.$$

Squaring both sides of the equation, we can obtain equation (1.27). On the contrary, if coordinate $M(x,y,z)$ satisfies the equation (1.27), there is always $|\overrightarrow{MC}|=R$, that is M on the sphere. So the equation (1.27) is a spherical equation, $C(x_0,y_0,z_0)$ is the center of the sphere, R is the radius of the sphere.

In particular, coordinate origin $O(0,0,0)$ is the center of the sphere, R is the radius of the sphere, its spherical equation is
$$x^2+y^2+z^2=R^2. \qquad (1.28)$$

2. Ellipsoid

The surface determined by equation
$$\frac{x^2}{a^2}+\frac{y^2}{b^2}+\frac{z^2}{c^2}=1 \quad (a>0,b>0,c>0) \qquad (1.29)$$
is called **ellipsoid**, and a,b,c are usually called half axes of the ellipsoid. The equation (1.28) is called the **standard equation of ellipsoid**.

The properties and figures of ellipsoid are discussed below.

1) Range of Figures

By equation (1.29), there are obviously
$$-a\leqslant x\leqslant a, \quad -b\leqslant y\leqslant b, \quad -c\leqslant z\leqslant c.$$
Thus, the ellipsoid is surrounded by a rectangular body surrounded by six planes, $x=\pm a, y=\pm b, z=\pm c$.

2) Symmetry

With $-x$ instead of x in equation (1.29), the equation is invariant, so the point (x,y,z) and the Oyz-plane symmetric point $(-x,y,z)$ are on the ellipsoid, that is, the ellipsoid is symmetric about the Oyz-plane. Similarly, the ellipsoid is also symmetric about the Ozx-plane and the Oxy-plane.

With $-x,-y$ instead of x,y in equation (1.29), the equation is invariant, the ellipsoid is shown to be symmetric about the z-axis. Similarly, the ellipsoid is also symmetric about the y-axis, the x-axis.

With $-x,-y,-z$ instead of x,y,z in equation (1.29), the equation is invariant, so the ellipsoid is symmetric about the origin.

The six intersection points $(\pm a,0,0)$, $(0,\pm b,0)$, $(0,0,\pm c)$ of the ellipsoid (1.29) and the three coordinate axes are called the **vertex** of the ellipsoid.

3) Cross Section

A plane that is parallel to the coordinate plane is to intercept the surface, and the resulting intersection is called the **cross section** of the surface.

With a set of planar $z=h(|h|\leqslant c)$ parallel to the Oxy-plane to cut ellipsoid (1.29) and cross section equation is
$$\begin{cases} \dfrac{x^2}{a^2}+\dfrac{y^2}{b^2}+\dfrac{z^2}{c^2}=1, \\ z=h. \end{cases}$$

This set of cross sections is ellipse, and the larger the $|h|$, the smaller the ellipse. When $|h|=c$, the cross sections are reduced to two points $(0,0,c)$ and $(0,0,-c)$; when $h=0$, that is using the Oxy-plane to intercept the ellipsoid, we get the largest cross section.

The same result can be obtained by using the plane parallel to the Oyz-plane and the Ozx-plane to intercept the ellipsoid (1.29).

In summary, the shape of the ellipsoid can be obtained, as shown in Figure 1.18.

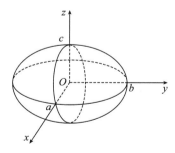

Figure 1.18
图 1.18

3. Hyperboloid

The hyperboloid is divided into hyperboloid of one sheet and hyperboloid of two sheets by the characteristics of the figure.

Determined by equation

$$\frac{x^2}{a^2}+\frac{y^2}{b^2}-\frac{z^2}{c^2}=1 \quad (a>0,b>0,c>0) \qquad (1.30)$$

is called the **hyperboloid of one sheet**.

Determined by equation

$$\frac{x^2}{a^2}+\frac{y^2}{b^2}-\frac{z^2}{c^2}=-1 \quad (a>0,b>0,c>0) \qquad (1.31)$$

is called the **hyperboloid of two sheets**.

The following is a graph of the hyperboloid of one sheet (1.30).

It is obvious that the hyperboloid of one sheet (1.30) is symmetric about the coordinate axis, coordinate plane and origin.

A set of planar $z=h$ parallel to the Oxy-plane is used to intercept a hyperboloid of one sheet, and the cross section is an ellipse, its equation is

$$\begin{cases} \dfrac{x^2}{a^2}+\dfrac{y^2}{b^2}=1+\dfrac{h^2}{c^2}, \\ z=h, \end{cases}$$

and the bigger the $|h|$, the bigger the ellipse.

With Oyz-plane truncated the hyperboloid of one sheet (1.30), obtain a hyperbola whose real axis is the y-axis.

With Ozx-plane truncated the hyperboloid of one sheet (1.30), obtain a hyperbola whose real axis is the x-axis.

Therefore, the figure of the hyperboloid of one sheet (1.30) is shown in Figure 1.19.

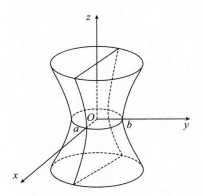

Figure 1.19
图 1.19

In the same way, the hyperboloid of two sheets

1.5 Quadratic Surfaces and Their Equations
1.5 二次曲面及其方程

(1.31) can also be obtained.

$z = h$ is used to cut the hyperboloid of two sheets (1.31), its cross section equation is

$$\begin{cases} \dfrac{x^2}{a^2} + \dfrac{y^2}{b^2} = \dfrac{h^2}{c^2} - 1, \\ z = h. \end{cases}$$

It is clear when $|h| < c$, there is no cross section; when $|h| = c$, the cross section is two points $(0, 0, \pm c)$; when $|h| > c$, the cross section is ellipse, and the bigger the $|h|$, the bigger the ellipse.

Cut the hyperboloid of two sheet (1.31) with the Oyz-plane. The cross section is a hyperbola whose real axis is z-axis.

Cut the hyperboloid of two sheet (1.31) with the Ozx-plane. The cross section is a hyperbola whose real axis is *y-axis*.

Therefore, the figure of the hyperboloid of two sheet (1.31) is shown in Figure 1.20.

(1.31)的图形.

用 $z = h$ 去截双叶双曲面(1.31)，截痕的方程为

$$\begin{cases} \dfrac{x^2}{a^2} + \dfrac{y^2}{b^2} = \dfrac{h^2}{c^2} - 1, \\ z = h. \end{cases}$$

可见，当 $|h| < c$ 时，无截痕；当 $|h| = c$ 时，截痕为两点$(0,0,\pm c)$；当 $|h| > c$ 时，截痕为椭圆，且 $|h|$ 越大，椭圆越大.

用 Oyz 平面去截双叶双曲面(1.31)，截痕是一条实轴为 z 轴的双曲线.

用 Ozx 平面去截双叶双曲面(1.31)，截痕是一条实轴为 z 轴的双曲线.

因此，双叶双曲面(1.31)的图形如图1.20所示.

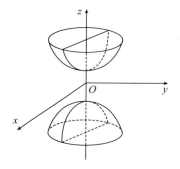

Figure 1.20
图 1.20

Notes (1) Equations
$$\dfrac{x^2}{a^2} - \dfrac{y^2}{b^2} + \dfrac{z^2}{c^2} = 1 \quad \text{and} \quad -\dfrac{x^2}{a^2} + \dfrac{y^2}{b^2} + \dfrac{z^2}{c^2} = 1$$
are also hyperboloid of one sheet.

(2) Equations
$$\dfrac{x^2}{a^2} - \dfrac{y^2}{b^2} + \dfrac{z^2}{c^2} = -1 \quad \text{and} \quad -\dfrac{x^2}{a^2} + \dfrac{y^2}{b^2} + \dfrac{z^2}{c^2} = -1$$
are also hyperboloid of two sheets.

注 （1）方程
$$\dfrac{x^2}{a^2} - \dfrac{y^2}{b^2} + \dfrac{z^2}{c^2} = 1 \quad 和 \quad -\dfrac{x^2}{a^2} + \dfrac{y^2}{b^2} + \dfrac{z^2}{c^2} = 1$$
也都是单叶双曲面；

（2）方程
$$\dfrac{x^2}{a^2} - \dfrac{y^2}{b^2} + \dfrac{z^2}{c^2} = -1 \quad 和 \quad -\dfrac{x^2}{a^2} + \dfrac{y^2}{b^2} + \dfrac{z^2}{c^2} = -1$$
也都是双叶双曲面.

4. Paraboloid

The common paraboloids contains elliptical paraboloid

4. 抛物面

常见的抛物面有椭圆抛物面和双曲抛

and hyperbolic paraboloid.

The surface determined by equation

$$z = \frac{x^2}{a^2} + \frac{y^2}{b^2} \quad (a>0, b>0, c>0) \tag{1.32}$$

is called **elliptic paraboloid**.

The surface determined by equation

$$z = \frac{x^2}{a^2} - \frac{y^2}{b^2} \quad (a>0, b>0, c>0) \tag{1.33}$$

is called **hyperbolic paraboloid**.

The graphs of elliptic paraboloid and hyperbolic paraboloid obtained by the cross section method are shown in Figures 1.21 and 1.22, respectively.

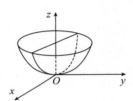

Figure 1.21

图 1.21

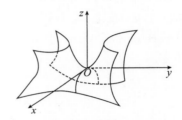

Figure 1.22

图 1.22

Remark A hyperbolic paraboloid is shaped like a saddle, so it is also called a **saddle surface**.

5. Cylinder

A surface formed by the translation of a straight line l along a curved Γ in space is called a **cylinder**. Moving straight line l is called the **generatrix** of a cylinder, fixed line Γ is called the directrix of a cylinder, as shown in Figure 1.23.

The common cylinders contain:

(1) **Cylindrical Surface**:
$$x^2 + y^2 = R^2 \quad \text{(Figure 1.24)}.$$

(2) **Elliptic Cylinder**:
$$\frac{x^2}{a^2} + \frac{y^2}{b^2} = 1 \quad \text{(Figure 1.25)}.$$

(3) **Hyperbolic Cylinder**:
$$\frac{y^2}{b^2} - \frac{x^2}{a^2} = 1 \quad (a>0, b>0) \quad \text{(Figure 1.26)}.$$

(4) **Paraboloid**:

$x^2 = 2py$ ($p>0$) (Figure 1.27).

Figure 1.23
图 1.23

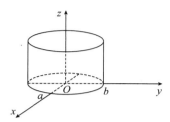

Figure 1.24
图 1.24

Figure 1.25
图 1.25

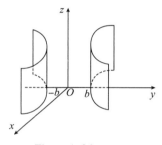

Figure 1.26
图 1.26

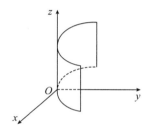

Figure 1.27
图 1.27

Remark If the surface equation is $F(x,y)=0$, then it must be a generatrix parallel to the z-axis, directrix is a curve Γ in Oxy-plane (in plane Cartesian coordinate system of equations is $F(x,y)=0$) of the cylinder. For example, cylinder $x^2+y^2=R^2$, it takes the circle on the Oxy-plane as the generatrix. A straight line parallel to the z-axis forms the cylinder of the directrix.

6. Rotating Surface

The surface formed by the rotation of a plane curve Γ around a fixed line l in the same plane is called a **rotating surface**. Γ is called the **generatrix** of rotation surface, fixed line l is called the **rotation axis** of a rotating surface, abbreviated as **axis**.

The spherical, cylindrical and other surfaces discussed previously are all rotating surfaces.

Theorem 1.6 The curve Γ is on the Oyz-plane, and

its plane Cartesian coordinate equation is
$$F(y,z)=0.$$
Then the rotating surface Σ of Γ rotates around the z-axis, and the Σ equation is
$$F(\pm\sqrt{x^2+y^2},z)=0. \quad (1.34)$$

Proof As shown in Figure 1.28, let $M(x,y,z)$ be any point on the rotating surface, and assume that the point M is obtained by turning the point $M_0(0,y_0,z_0)$, then $z=z_0$, and the distance from M to the z-axis is equal to the distance from M_0 to the z-axis. The distance from M and z-axis is $\sqrt{x^2+y^2}$, and the distance between M_0 and z-axis is $\sqrt{y_0^2}=|y_0|$. Therefore, $y_0=\pm\sqrt{x^2+y^2}$. And because M_0 is on Γ, so $F(y_0,z_0)=0$, substitute $z_0=z, y_0=\pm\sqrt{x^2+y^2}$, there is
$$F(\pm\sqrt{x^2+y^2},z)=0,$$
that is, the coordinates of the point $M(x,y,z)$ at the rotating surface satisfy the equation
$$F(\pm\sqrt{x^2+y^2},z)=0.$$

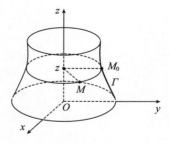

Figure 1.28
图 1.28

In contrast, if $M(x,y,z)$ coordinates satisfy the equation $F(\pm\sqrt{x^2+y^2},z)=0$, it is easy to prove that M is on rotating surface Σ.

Then the equation of the rotating surface Σ is
$$F(\pm\sqrt{x^2+y^2},z)=0.$$

Remark This theorem states that if the generatrix Γ of rotating surfaces is on the Oyz-plane, its plane Cartesian coordinatesystem equation is $F(y,z)=0$. To write the

equation for the rotating surface of a curve Γ that rotates around the z-axis, just change the y in the equation $F(y,z)=0$ to $\pm\sqrt{x^2+y^2}$. Similarly, the equation for the rotating surface of a curve Γ that rotates around the y-axis is
$$F(y,\pm\sqrt{x^2+z^2})=0,$$
that is, change the z in the equation $F(y,z)=0$ to $\pm\sqrt{x^2+z^2}$. Similar results have been obtained for the cases where the curves on the other coordinate planes rotate about the axes.

In addition, whether an equation represents a rotating surface depends on whether the equation contains the sum of squares of the two variables.

Example 1 Find the equation of the rotating surface formed by the rotation of the hyperbola $\dfrac{x^2}{9}-\dfrac{y^2}{4}=1$ on the Oxy-plane around the x-axis.

Solution Because of the rotation of the x-axis, it is only necessary to change the y in the equation $\dfrac{x^2}{9}-\dfrac{y^2}{4}=1$ to $\pm\sqrt{y^2+z^2}$, so the equation of the rotating surface is
$$\frac{x^2}{9}-\frac{y^2+z^2}{4}=1.$$
The curved surface is a **rotating hyperboloid of two sheets**.

Example 2 Find the following equations for the rotating surface:

(1) The rotating surface formed by the circle $\dfrac{x^2}{b^2}+\dfrac{y^2}{a^2}=1(b>0,a>0)$ on the Oxy-plane rotated around the x-axis and the y-axis;

(2) The rotating surface formed by the parabola $x^2=az$ $(a>0)$ on the Ozx-plane rotated around the symmetry-axis;

(3) The rotating surface formed by the hyperbola $-\dfrac{y^2}{b^2}+\dfrac{z^2}{a^2}=1$ on the Oyz-plane rotated around the real axis and the imaginary-axis;

(4) The rotating surface formed by the line $y=ax+b$ on the Oxy-plane rotated around the x-axis and the y-axis.

所形成的旋转曲面的方程,则只需将方程 $F(y,z)=0$ 中的 y 换成 $\pm\sqrt{x^2+y^2}$ 即可. 同理,曲线 Γ 绕 y 轴旋转所形成的旋转曲面的方程为
$$F(y,\pm\sqrt{x^2+z^2})=0,$$
即将 $F(y,z)=0$ 中的 z 换成 $\pm\sqrt{x^2+z^2}$. 对于其他坐标平面上的曲线绕坐标轴旋转的情形,有类似的结果.

另外,一个方程是否表示旋转曲面,只需看方程中是否含有两个变量的平方和.

例 1 求 Oxy 平面上的双曲线 $\dfrac{x^2}{9}-\dfrac{y^2}{4}=1$ 绕 x 轴旋转所形成的旋转曲面的方程.

解 由于绕 x 轴旋转,只需将方程
$$\frac{x^2}{9}-\frac{y^2}{4}=1$$
中的 y 换成 $\pm\sqrt{y^2+z^2}$ 即可,故所求的旋转曲面方程为
$$\frac{x^2}{9}-\frac{y^2+z^2}{4}=1.$$
该曲面称为**旋转双叶双曲面**.

例 2 求出下列旋转曲面的方程:

(1) 由 Oxy 平面上的椭圆 $\dfrac{x^2}{b^2}+\dfrac{y^2}{a^2}=1$ $(b>0,a>0)$ 绕 x 轴和 y 轴旋转所形成的旋转曲面;

(2) Ozx 平面上的抛物线 $x^2=az$ $(a>0)$ 绕其对称轴旋转所形成的旋转曲面;

(3) Oyz 平面上的双曲线 $-\dfrac{y^2}{b^2}+\dfrac{z^2}{a^2}=1$ $(b>0,a>0)$ 绕其实轴和虚轴旋转所形成的旋转曲面;

(4) Oxy 平面上的直线 $y=ax+b$ 绕 x 轴和 y 轴旋转所形成的旋转曲面.

Solution (1) The rotating surface equation rotating around the x-axis is
$$\frac{x^2}{b^2}+\frac{y^2+z^2}{a^2}=1.$$
Rotating around the y-axis, the rotating surface equation is
$$\frac{x^2+z^2}{b^2}+\frac{y^2}{a^2}=1. \qquad (1.35)$$
The above surfaces are called the **rotating elliptical surfaces** (Figure 1.29 shows rotating elliptical surface (1.35)).

(2) The rotating surface equation rotating around the symmetric axis (z-axis) is
$$x^2+y^2=az.$$
This surface is called a **rotating paraboloid** (Figure 1.30).

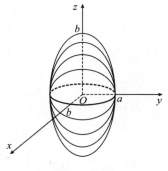

Figure 1.29

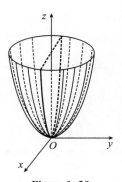

Figure 1.30

(3) The rotating surface equation rotating around the real axis (z-axis) is
$$-\frac{x^2+y^2}{b^2}+\frac{z^2}{a^2}=1.$$
The surface is called a **rotating hyperboloid of two sheets** (Figure 1.31). The rotating surface equation rotating around the imaginary-axis (y-axis) is
$$-\frac{y^2}{b^2}+\frac{x^2+z^2}{a^2}=1.$$
The surface is called a **rotating hyperboloid of one sheet** (Figure 1.32).

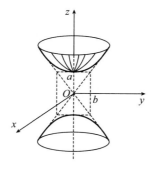

Figure 1.31
图 1.31

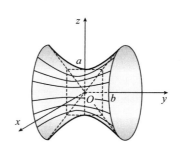

Figure 1.32
图 1.32

(4) The rotating surface equation rotating around the x-axis is
$$\pm\sqrt{y^2+z^2}=ax+b,$$
that is
$$y^2+z^2=(ax+b)^2,$$
which is the **conical surface** of the vertex at $\left(-\dfrac{b}{a},0,0\right)$. The rotating surface rotating around the y-axis is
$$y=\pm a\sqrt{x^2+z^2}+b,$$
that is
$$(y-b)^2=a^2(x^2+z^2),$$
which is the conical surface of the vertices at $(0,b,0)$.

In particular, if $b=0$, that is, the generatrix is the line $y=ax$ passes through the origin, then the vertices of the cone surface rotates around the x-axis or y-axis are at the origin, and the equations are
$$a^2x^2=y^2+z^2 \quad \text{and} \quad y^2=a^2(x^2+z^2)$$
respectively.

（4）绕 x 轴旋转所得旋转曲面的方程为
$$\pm\sqrt{y^2+z^2}=ax+b,$$
即
$$y^2+z^2=(ax+b)^2,$$
它表示顶点在 $\left(-\dfrac{b}{a},0,0\right)$ 的**圆锥面**；绕 y 轴旋转所得旋转曲面的方程为
$$y=\pm a\sqrt{x^2+z^2}+b,$$
即
$$(y-b)^2=a^2(x^2+z^2),$$
它是顶点在 $(0,b,0)$ 的圆锥面.

特别地，若 $b=0$，即母线为经过原点的直线 $y=ax$，则绕 x 轴或 y 轴旋转而成的圆锥面的顶点都在原点，方程分别为
$$a^2x^2=y^2+z^2 \quad \text{和} \quad y^2=a^2(x^2+z^2).$$

1.6　Space Curves and Their Equations
1.6　空间曲线及其方程

This section mainly introduces the concept of space curve equation and several different forms.

本节主要介绍空间曲线方程的概念及几种不同形式的曲线方程.

1. General Equation of Space Curve

As we known, a straight line is the intersection of two planes, and their equations are the equations of simultaneous two plane equations. Similarly, space curve can be regarded

1. 空间曲线的一般方程

我们知道，直线就是两个平面的交线，它的方程就是联立两个平面方程的方程组. 类似地，空间中的曲线可以看作两个曲面的交线.

as an intersecting line of two curved surfaces.

Let the equation of the surface Σ_1 is $F_1(x,y,z)=0$, and the equation of the surface Σ_2 is $F_2(x,y,z)=0$. The intersection of them is Γ (Figure 1.33), then the coordinate of an arbitrary point $M(x,y,z)$ on the curve Γ satisfies the equations of the two surfaces at the same time, that is

$$\begin{cases} F_1(x,y,z)=0, \\ F_2(x,y,z)=0. \end{cases} \quad (1.36)$$

Conversely, the points that are solutions of equations (1.36) are on surfaces Σ_1 and Σ_2 at the same time, so that they are on the intersection Γ of surfaces Σ_1 and Σ_2. Therefore, equations (1.36) is a **general equation** of Γ.

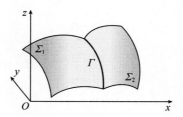

Figure 1.33
图 1.33

2. Parametric Equation of Space Curve

A curve is essentially a one-dimensional graph, that is, for any point on the curve, if a coordinate is determined, the other two coordinates are determined, that is, it has only one degree of freedom. This essentially determines that if its equation is represented by a parameter, then there is only one parameter. Therefore, the general form of parametric parametric equation should be

$$\begin{cases} x=x(t), \\ y=y(t), \quad (\alpha \leqslant t \leqslant \beta). \\ z=z(t) \end{cases} \quad (1.37)$$

In fact, the coordinate x,y,z of the moving point M on space curve could be expressed as a function of the parameter, thus having the form as equation (1.37). Conversely, given $t \in [\alpha,\beta]$, one point on the curve could be obtained from equation (1.37), and the whole curve is obtained when t gets through $[\alpha, \beta]$.

3. The Projection of a Space Curve on a Coordinate Surface

1) A Projective Curve of a Space Curve on a Coordinate Surface

For the general space curve Γ, make a cylindrical Σ_z with directrix Γ whose generatrix line is parallel to the z-axis, we call the intersection of Σ_z and Oxy-plane C_z the **projective curve** of Γ in the Oxy-plane (referred to as **projection**), and call the cylindrical Σ_z the **projective cylinder** of Γ about Oxy-plane (Figure 1.34).

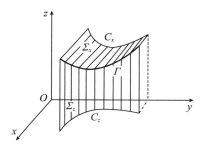

Figure 1.34

Similarly, if the generatrix of the cylinder is parallel to the x-axis or the y-axis, we can obtain the projection of Γ on the Oyz-plane or on the Ozx-plane C_x, C_y, and the corresponding projective cylinder Σ_x, Σ_y.

The projection curve on the coordinate plane not only can enhance the visual image of the curve, but also help to understand the variation range of the curve.

2) The Equation of the Projection Curve Obtained from the General Equation of the Space Curve

In order to find the equation of the projective C_z of the space curve Γ on the Oxy-plane, Γ can be expressed as equations:

$$\begin{cases} f(x,y) = 0, \\ g(x,y,z) = 0. \end{cases} \quad (1.38)$$

The equation $f(x,y) = 0$ represents the cylinder Σ_z whose generatrix is parallel to the z-axis. This represents

the curve Γ as the intersection of Σ_z with another surface $g(x,y,z)=0$. Thus Σ_z is exactly the projective cylinder of Γ with respect to Oxy-plane, so
$$\begin{cases} f(x,y)=0, \\ z=0 \end{cases}$$
is the equation of the projection C_z of Γ in the Oxy-plane.

If the general equation of Γ is
$$\begin{cases} F(x,y,z)=0, \\ G(x,y,z)=0. \end{cases} \tag{1.39}$$

In order to find the equation of projection C_z in the Oxy-plane, we make the equivalent transformation of the (1.39) to the form (1.38) by eliminating z.

Similarly, if x or y is eliminated in one of the two equations (1.39), it becomes a form
$$\begin{cases} f(x,z)=0, \\ g(x,y,z)=0. \end{cases} \quad \text{or} \quad \begin{cases} f(y,z)=0, \\ g(x,y,z)=0 \end{cases}$$
So
$$\begin{cases} f(y,z)=0, \\ x=0 \end{cases} \quad \text{and} \quad \begin{cases} f(x,z)=0, \\ y=0 \end{cases}$$
are the projection C_x of Γ in the Oyz-plane and the projection C_y of Γ in the Ozx-plane, respectively.

Example 1 What kind of curve will be represented by the equation $\begin{cases} x^2+y^2+z^2=25, \\ z=3 \end{cases}$?

Solution The equations represent the radius of the spherical surface 5: intersection of the $x^2+y^2+z^2=5^2$ which center is at origin and radius is 5, and the plane $z=3$, it is a circle on the plane $z=3$, whose center is $(0,0,3)$ and radius is 4 (Figure 1.35).

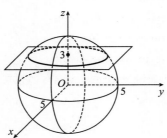

Figure 1.35
图 1.35

Example 2 Find the intersection line of the spherical $x^2+y^2+z^2=(2R)^2$ and the cylinder $(x-R)^2+y^2=R^2$, and make its sketch.

Solution The equation of the required intersection curve is
$$\begin{cases} x^2+y^2+z^2=(2R)^2, \\ (x-R)^2+y^2=R^2. \end{cases}$$
This curve is the intersection of a cylinder and a spherical surface. The cylindrical surface passes the center of the sphere, and the diameter of the cylindrical's top surface equals to the radius of the sphere. Suppose the sphere center as the origin and its radius is $2R$, then the curve we are going to find out is shown in Figure 1.36 (In the figure only half of the curve is shown). This required intersection in mathematics is often called the **Viviani curve**.

例 2 求球面 $x^2+y^2+z^2=(2R)^2$ 与圆柱面 $(x-R)^2+y^2=R^2$ 的交线,并作出其草图.

解 所求交线的方程为
$$\begin{cases} x^2+y^2+z^2=(2R)^2, \\ (x-R)^2+y^2=R^2. \end{cases}$$
这条曲线是圆柱面与球面的交线,其中圆柱面过球心且其直径与球面的半径相等,球面的球心为坐标原点,半径为 $2R$. 所以所求交线的图形如图 1.36 所示(图中仅画出了上半球面上的交线 Γ). 这条交线在数学上通常称为**维维安尼曲线**.

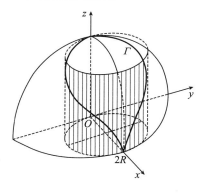

Figure 1.36
图 1.36

Example 3 On a transparent rectangular paper, there is a straight line l with an angle of θ to the base, and now it is rolled into a cylinder of R radius(Figure 1.36). If the thickness of the paper is neglected, the rectangle becomes a straight circle cylinder, and the l becomes a curve on the cylindrical surface of the winding. This curve is called an **isometric spiral**. It is characterized by that the distance between two adjacent circles is equal to $b=2\pi R\tan\theta$, and θ is called a **spiral angle**, b is called the **pitch**. Try to find the equation of isometric spiral.

Solution Establish the coordinate system as shown in Figure 1.37, where the x-axis passes through the intersection

例 3 设在一张透明的矩形纸上有一条与底边成 θ 角的直线 l. 现在把它卷成半径为 R 的圆筒(图 1.36). 若忽略纸的厚度,则矩形纸成为圆柱面,l 成为圆柱面上的曲线. 称此曲线为**等距螺线**,它的特征是相邻两圈之间等距,距离为 $b=2\pi R\tan\theta$. 通常称 θ 为**螺旋角**,称 b 为**螺距**. 试求等距螺线的方程.

解 建立直角坐标系如图 1.37 所示,其中 x 轴经过直线 l 与矩形底边交点. 任取等

of the L and the rectangular base. The $M(x,y,z)$ is used as the point on the spiral line, and the projection of M on the Oxy-plane is M_1, and the angle from the x-axis to the OM_1 is φ, then

$$z = M_1M = \frac{\varphi}{2\pi}b = (R\tan\theta)\varphi,$$
$$x = R\cos\varphi, \qquad (1.40)$$
$$y = R\sin\varphi.$$

Conversely, we know by the inverse process, if the coordinates of the point $M(x,y,z)$ satisfy the equation (1.40), then the M must be on the isometric spiral line, thus the equation of the isometric spiral is obtained

$$\begin{cases} x = R\cos\varphi, \\ y = R\sin\varphi, \\ z = (R\tan\theta)\varphi \end{cases} (\varphi \geq 0).$$

The resulting equation contains a parameter φ, hence it is called the **parametric equation of isometric spirals**.

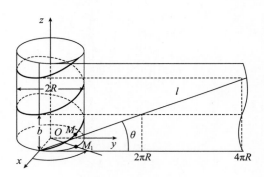

Figure 1.37

图 1.37

Example 4 Find the curve Γ represented by the parametric equation

$$\begin{cases} x = \cos t + \sin t, \\ y = \cos t - \sin t, \\ z = 1 - \sin 2t. \end{cases}$$

Solution The sum of squares on the two sides of the first two equations is added as

$$x^2 + y^2 = 2,$$

and

$$y^2 = 1 - 2\cos t \sin t = 1 - \sin 2t = z,$$

so the equation of the curve Γ can be written as
$$\begin{cases} x^2 + y^2 = 2, \\ y^2 = z. \end{cases}$$

Therefore, the curve represented by the parametric equation is the intersection of the cylindrical surface $x^2 + y^2 = 2$ and the parabolic cylinder $y^2 = z$, and its image is shown in Figure 1.38.

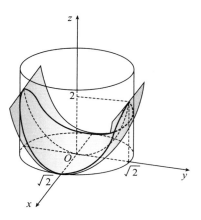

Figure 1.38

Example 5 Find the projection curve C_z of Γ on the Oxy-plane and the projection curve C_x of Γ on the Oyz plane, which Γ is the intersection of the surface $4z = 2x^2 + y^2$ and the plane $x - z = 0$.

Solution The equation of intersection Γ is
$$\begin{cases} x - z = 0, \\ 4z = 2x^2 + y^2. \end{cases} \quad (1.41)$$

In order to obtain the equation of projection curve C_z, we should eliminate z from two equations of (1.41). For this reason, the first equation of (1.41) $z = x$ is substituted into the second equation of (1.41), we get
$$4x = 2x^2 + y^2,$$
that is
$$(x-1)^2 + \frac{y^2}{2} = 1.$$

Thus the equation of projection curve C_z is

$$\begin{cases} (x-1)^2 + \dfrac{y^2}{2} = 1, \\ z = 0. \end{cases}$$

This is an ellipse on the Oxy-plane (Figure 1.39).

In order to obtain the equation of projection curve C_x, we should eliminate x from two equations of (1.41). For this reason, the first equation of (1.41) $x = z$ is substituted into the second equation of (1.41), we get
$$4z = 2z^2 + y^2,$$
that is
$$(z-1)^2 + \dfrac{y^2}{2} = 1.$$

Thus the equation of C_x is
$$\begin{cases} (z-1)^2 + \dfrac{y^2}{2} = 1, \\ x = 0. \end{cases}$$

This is an ellipse on the Oyz-plane (Figure 1.39).

$$\begin{cases} (x-1)^2 + \dfrac{y^2}{2} = 1, \\ z = 0. \end{cases}$$

这是 Oxy 平面上的一个椭圆(图 1.39).

为了求得投影曲线 C_x 的方程,应该在方程组(1.41)的两个方程中消去 x. 为此,把由第一个方程得到的 $x = z$ 代入第二个方程,得
$$4z = 2z^2 + y^2,$$
即
$$(z-1)^2 + \dfrac{y^2}{2} = 1.$$

由此可得 C_x 的方程为
$$\begin{cases} (z-1)^2 + \dfrac{y^2}{2} = 1, \\ x = 0. \end{cases}$$

这是 Oyz 平面上的一个椭圆(图 1.39).

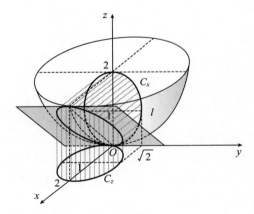

Figure 1.39

图 1.39

Exercises 1
习题 1

1. Let vector $u = 2a + 3b - c$, $v = a - 2b + 5c$, try to represent $3u - 4v$ with the vector a, b, c.

1. 设向量 $u = 2a + 3b - c$, $v = a - 2b + 5c$, 试用 a, b, c 表示 $3u - 4v$.

Exercises 1

2. Prove that the triangle with the vertex of $M_1(4,3,1)$, $M_2(7,1,2)$, $M_3(5,2,3)$ is an isosceles triangle.

3. Find the distance from the point $P(2,-6,4)$ to the origin, the coordinate axis and the coordinate plane.

4. Let point P be on the x-axis, the distance to $P_1(0,\sqrt{2},3)$ is two times to $P_2(0,1,-1)$, find coordinate of point P.

5. Let $a=i+2j+3k, b=2i-2j+3k$, find
 (1) $a+b$; (2) $a-b$; (3) $2a-3b$.

6. Find λ that makes vector $a=\{\lambda,2,10\}$ parallel to vector $b=\{2,10,50\}$.

7. Let two points $A(1,-1,2), B(4,1,3)$, find
 (1) the coordinates and the partial vectors of $\overrightarrow{AB}$ on the three axes;
 (2) the direction cosine of vector $\overrightarrow{AB}$.

8. Two points $A(2,\sqrt{2},5), B(3,0,6)$ are known, find the mode, the direction cosine and the direction angle of the vector $\overrightarrow{AB}$.

9. It is known that $a=\{-1,3,2\}, b=\{2,5,-1\}, c=\{6,4,-6\}$, prove that $a-b$ is parallel to c.

10. It is known that the vectors $a=\{1,0,1\}, b=\{0,1,0\}, c=\{0,0,1\}$, find
 (1) $a \cdot b, a \cdot c, b \cdot c$;
 (2) $a \times a, a \times b, a \times c, b \times c$.

11. It is known that the vectors $a=\{1,0,0\}, b=\{2,2,1\}$, find angle cosine between $a \cdot b$ and $a \times b$, a and b.

12. Let the vectors $a=-i+3j+4k, b=5i+2j-k$, calculate the angle cosine between
 (1) $a \cdot b$ and $a \times b$;
 (2) $(-2a) \cdot 3b$ and $a \times 2b$;
 (3) a and b.

13. It is known that $|a|=5, |b|=2, \langle a,b \rangle = \dfrac{\pi}{3}$, find the modulus of $c=2a-3b$.

14. Let vector $a=i+3j-k, b=2j+3k$, calculate $a \times b$, and the area of the parallelogram with a and b as adjacent sides.

习题 1

2. 求证：以 $M_1(4,3,1), M_2(7,1,2), M_3(5,2,3)$ 三点为顶点的三角形是一个等腰三角形.

3. 求点 $P(2,-6,4)$ 到原点及各坐标轴和各坐标面的距离.

4. 设点 P 在 x 轴上，它到点 $P_1(0,\sqrt{2},3)$ 的距离为到点 $P_2(0,1,-1)$ 的距离的两倍，求点 P 的坐标.

5. 设 $a=i+2j+3k, b=2i-2j+3k$，求：
 (1) $a+b$；(2) $a-b$；(3) $2a-3b$.

6. 求 λ，使向量 $a=\{\lambda,2,10\}$ 与 $b=\{2,10,50\}$ 平行.

7. 设有两点 $A(1,-1,2), B(4,1,3)$，求：
 (1) $\overrightarrow{AB}$ 的坐标和在三条坐标轴上的分向量；
 (2) $\overrightarrow{AB}$ 的方向余弦.

8. 已知两点 $A(2,\sqrt{2},5), B(3,0,6)$，求向量 $\overrightarrow{AB}$ 的模、方向余弦和方向角.

9. 已知向量 $a=\{-1,3,2\}, b=\{2,5,-1\}, c=\{6,4,-6\}$，证明：$a-b$ 与 c 平行.

10. 已知向量 $a=\{1,0,1\}, b=\{0,1,0\}, c=\{0,0,1\}$，求：
 (1) $a \cdot b, a \cdot c, b \cdot c$；
 (2) $a \times a, a \times b, a \times c, b \times c$.

11. 已知向量 $a=\{1,0,0\}, b=\{2,2,1\}$，求 $a \cdot b, a \times b$ 以及 a 与 b 的夹角余弦.

12. 设向量 $a=-i+3j+4k, b=5i+2j-k$，计算：
 (1) $a \cdot b$ 及 $a \times b$；
 (2) $(-2a) \cdot 3b$ 及 $a \times 2b$；
 (3) a 与 b 的夹角余弦.

13. 已知 $|a|=5, |b|=2, \langle a,b \rangle = \dfrac{\pi}{3}$，求 $c=2a-3b$ 的模.

14. 设向量 $a=i+3j-k, b=2j+3k$，计算 $a \times b$，并计算以 a,b 为邻边的平行四边形的面积.

15. Find vector **b**, co-line with vector $\mathbf{a}=\{2,-1,2\}$ and satisfied $\mathbf{a}\cdot\mathbf{b}=-18$.

16. Point out the special positions of the following planes and draw out the planes:
(1) $x=0$;　　　　(2) $3y-1=0$;
(3) $2x-7y-6=0$;　(4) $x-5y=0$;
(5) $y+z=1$;　　(6) $x-2z=0$;
(7) $6x+5y-z=0$.

17. Find a plane that pass point $M(1,3,4)$ and take $\mathbf{n}=\{2,2,1\}$ as the normal vector of the plane.

18. Find a plane that pass three points $A(1,0,0)$, $B(0,1,0)$, $C(0,0,1)$.

19. Find a plane equation which passes the original point and point $M(1,1,-1)$, and is perpendicular to the plane $4x+3y+z-1=0$.

20. Find a plane which passes $(0,0,1)$ and parallel to the plane
$$3x+4y+2z=1.$$

21. Find a plane that passes x-axis and the point $(4,-3,-1)$.

22. Find a straight line equation that passes the point $M(1,0,-2)$ and perpendicular to the two lines
$$\frac{x-1}{1}=\frac{y}{1}=\frac{z+1}{-1},\quad \frac{x}{1}=\frac{y-1}{-1}=\frac{z+1}{0}.$$

23. Find a plane equation which passes the point $M(2,-3,-5)$ and perpendicular to the plane
$$6x-3y-5z+2=0.$$

24. Find a plane equation which passes the point $(1,2,3)$ and a line
$$\frac{x-4}{-1}=\frac{y+3}{4}=\frac{z}{-2}.$$

25. Find a plane equation which passes the point $M(2,1,0)$ and perpendicular to the line
$$\begin{cases} x=2t-3,\\ y=3t+5,\\ z=t. \end{cases}$$

26. Find the angle between line
$$\frac{x-1}{3}=\frac{y+1}{4}=\frac{z-1}{5}$$

and line
$$\frac{x}{-1}=\frac{y+1}{2}=\frac{z}{2}.$$

27. Determine the relationship between the line l and the plane π in the following groups:

(1) $l: \frac{x-3}{-2}=\frac{y+4}{7}=\frac{z}{3}$,

$\pi: 4x-2y-2z=3$;

(2) $l: \begin{cases} 5x-3y+2z-5=0, \\ 2x-y-z-1=0, \end{cases}$

$\pi: 4x-3y+7z-7=0$;

(3) $l: \begin{cases} x=t, \\ y=-2t+9, \\ z=9t-4, \end{cases}$

$\pi: 3x-4y+7z-10=0$.

28. Determine the positional relationship between line
$$l_1: \frac{x}{0}=\frac{y}{3}=\frac{z-6}{3}$$
and line
$$l_2: \begin{cases} x+y+z-3=0, \\ y+z=2. \end{cases}$$

29. Determine the location of the following line l and plane π, and if they intersect, find the intersection point coordinates.

(1) $\pi: 2x+y-z=0$,

$l: \frac{x-1}{1}=\frac{y+1}{1}=\frac{z}{3}$;

(2) $\pi: x-y+2z=0$,

$l: \begin{cases} x+y+z=1, \\ 2x+3z=1; \end{cases}$

(3) $\pi: x+2y-5z-11=0$,

$l: \frac{x-5}{2}=\frac{y+3}{-2}=\frac{z-1}{3}$.

30. Find the following equations of rotational surfaces:

(1) The rotating surface is generated by the parabolic $y^2=5x$ on the Oxy-plane rotating around the x-axis.

(2) The rotating surface is generated by the hyperbolic

$\frac{x^2}{a^2} - \frac{z^2}{c^2} = 1$ on the Ozx-plane respectively rotating around the x-axis and the z-axis.

31. Point out the following surfaces, which are rotating surfaces, and if they are rotational surfaces, show how they are generated?

   (1) $x^2 + y^2 + z^2 = 1$;

   (2) $x^2 + 2y^2 + 3z^2 = 1$;

   (3) $\frac{x^2}{9} + \frac{y^2}{4} + \frac{z^2}{9} = 1$;

   (4) $x^2 - \frac{y^2}{4} + z^2 = 1$.

32. Indicate what the following equations represent in the Cartesian coordinate system and the space Cartesian coordinate system:

   (1) $y = 5x$;   (2) $\frac{x^2}{9} + \frac{y^2}{16} = 1$;

   (3) $x^2 - y^2 = 1$;   (4) $y^2 = 4x$.

33. Indicate the names of the following surfaces:

   (1) $\frac{x^2}{4} + \frac{z^2}{25} = 1$;

   (2) $y^2 = 2z$;

   (3) $x^2 + y^2 + z^2 - 2x = 1$;

   (4) $\frac{x^2}{4} + \frac{y^2}{3} + \frac{z^2}{3} = 1$.

34. Indicate the shape of the curves represented by the following equations:

   (1) $\begin{cases} 3x^2 + y^2 = z, \\ y = 3; \end{cases}$

   (2) $\begin{cases} x^2 + 4y^2 + 9z^2 = 30, \\ z = 1; \end{cases}$

   (3) $\begin{cases} x^2 - 4y^2 + z^2 = 25, \\ x = -3; \end{cases}$

   (4) $\begin{cases} y^2 + z^2 - 4x + 8 = 0, \\ y = 4; \end{cases}$

   (5) $\begin{cases} \frac{y^2}{9} - \frac{z^2}{4} = 1, \\ x - 2 = 0. \end{cases}$

35. Find the projection curve equation of the curve

$$\begin{cases} z = 2 - x^2 - y^2, \\ z = (x-1)^2 + (y-1)^2 \end{cases}$$ on three coordinate planes.

36. Find the projection of the cylinder $z = \sqrt{x^2 + y^2}$ and a cylinder $z^2 = 2x$ to form a solid into three coordinate planes.

37. Find the projection of the solid surrounded by the surface $z = \sqrt{6 - x^2 - y^2}$ and $x^2 + y^2 = z$ on the Oxy-plane.

38. A solid is surrounded by an upper hemispherical $z = \sqrt{4 - x^2 - y^2}$ and a cone $z = \sqrt{3(x^2 + y^2)}$, find its projection on the Oxy-plane.

39. Find the Cartesian coordinate equation of the projection of a helical line
$$\begin{cases} x = a\cos\theta, \\ y = a\sin\theta, \\ z = b\theta \end{cases}$$
on three coordinate planes.

40. Translate the curve equation
$$\begin{cases} x = a\cos^2 t, \\ y = a\sin^2 t, \quad (0 \leqslant t \leqslant 2\pi) \\ z = a\sin 2t \end{cases}$$
into a general equation.

36. 求锥面 $z = \sqrt{x^2 + y^2}$ 与柱面 $z^2 = 2x$ 所围成的立体在三个坐标面上的投影.

37. 求曲面 $z = \sqrt{6 - x^2 - y^2}$ 与 $x^2 + y^2 = z$ 所围成的立体在 Oxy 平面上的投影.

38. 设一个立体由上半球面 $z = \sqrt{4 - x^2 - y^2}$ 和锥面 $z = \sqrt{3(x^2 + y^2)}$ 所围成,求它在 Oxy 平面上的投影.

39. 求螺旋线
$$\begin{cases} x = a\cos\theta, \\ y = a\sin\theta, \\ z = b\theta \end{cases}$$
在三个坐标面上的投影曲线的直角坐标方程.

40. 化曲线方程
$$\begin{cases} x = a\cos^2 t, \\ y = a\sin^2 t, \quad (0 \leqslant t \leqslant 2\pi) \\ z = a\sin 2t \end{cases}$$
为一般方程.

Chapter 2   Derivatives for the Function of Several Variables
第 2 章   多元函数的微分

In this chapter, we will study the function of several variable and its differential calculus. Among it, many concepts are extension of differential calculus of univers function.

在本章中,我们将学习多元函数及其微分学,其中许多概念是我们学过的一元函数微分学知识的延伸.

2.1   The Basic Concept of the Function of Several Variables
2.1   多元函数的基本概念

1. Planar Point Set

We usually use
$$\mathbf{R}^2 = \{(x,y) \mid x,y \in \mathbf{R}\}$$
to represent the whole Oxy-plane. While the symbol $E \subset \mathbf{R}^2$ indicates that E is a set of plane points, and the symbol $P_0 \in E$ indicates that P_0 is a point in E. We will introduce some terms relative to the planar point set.

Neighborhood: Let $P_0(x_0, y_0)$ be a point in the Oxy-plane, $\delta > 0$, the whole points of which the distance from P_0 are less than δ are called the δ **neighborhood** of P_0, written as $U(P_0, \delta)$, that is
$$U(P_0, \delta) = \{(x,y) \mid \sqrt{(x-x_0)^2 - (y-y_0)^2} < \delta\}.$$

As shown in Figure 2.1(a), $U(P_0, \delta)$ is a circle on Oxy-plane which take P_0 as its center and δ as its radius, excluding the circumference.

Deleted neighborhood: The δ deleted neighborhood of P_0 is recorded as $\overset{\circ}{U}(P_0, \delta)$, and defined as
$$\overset{\circ}{U}(P_0, \delta) = \{(x,y) \mid 0 < \sqrt{(x-x_0)^2 - (y-y_0)^2} < \delta\}.$$

It can be seen that the δ deleted neighborhood $\overset{\circ}{U}(P_0, \delta)$ of P_0 is the δ neighborhood $U(P_0, \delta)$ of P_0 which

1. 平面点集

我们通常用
$$\mathbf{R}^2 = \{(x,y) \mid x,y \in \mathbf{R}\}$$
表示整个 Oxy 平面,而用符号 $E \subset \mathbf{R}^2$ 表示 E 是平面点集,用符号 $P_0 \in E$ 表示 P_0 是 E 中的一点.下面介绍几个与平面点集有关的术语.

邻域:设 $P_0(x_0, y_0)$ 是 Oxy 平面上的一个点,$\delta > 0$,与点 P_0 距离小于 δ 的点的全体称为点 P_0 的 δ **邻域**,记为 $U(P_0, \delta)$,即
$$U(P_0, \delta)$$
$$= \{(x,y) \mid \sqrt{(x-x_0)^2 - (y-y_0)^2} < \delta\}.$$

如图 2.1(a)所示,$U(P_0, \delta)$ 就是 Oxy 平面上以点 P_0 为圆心,δ 为半径的圆的内部,不包括圆周.

去心邻域:点 P_0 的 δ **去心邻域**记为 $\overset{\circ}{U}(P_0, \delta)$,定义为
$$\overset{\circ}{U}(P_0, \delta)$$
$$= \{(x,y) \mid 0 < \sqrt{(x-x_0)^2 - (y-y_0)^2} < \delta\}.$$

可见,P_0 的 δ 去心邻域 $\overset{\circ}{U}(P_0, \delta)$ 就是 P_0 的 δ 邻域 $U(P_0, \delta)$ 去掉点 P_0(图 2.1(b)).在不

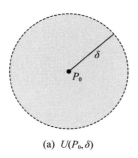

(a) $U(P_0,\delta)$ (b) $\overset{\circ}{U}(P_0,\delta)$

Figure 2.1
图 2.1

is removed P_0 (Figure 2.1(b)). When it is not necessary to emphasize the size of δ, $U(P_0,\delta)$ and $\overset{\circ}{U}(P_0,\delta)$ are often written as $U(P_0)$ and $\overset{\circ}{U}(P_0)$, respectively.

Interior point: If there is a neighborhood $U(P,\delta)$ of P satisfied $U(P,\delta) \subset E$, we say P is an **interior point** of a set E.

For example in Figure 2.2, the point P_1 is an interior point of set E.

Exterior point: If there is a neighborhood $U(P,\delta)$ of P satisfied $U(P,\delta) \cap E = \varnothing$, we say P is an **exterior point** of set E.

For example, in Figure 2.2, the point P_2 is an exterior point of a set E.

Boundary point: P is a **boundary point** of E if every neighborhood of P contains points that are in E and points that are not in E.

For example, in Figure 2.2, P_3 is the boundary point of set E.

需要强调 δ 的大小时,常常将 $U(P_0,\delta)$ 和 $\overset{\circ}{U}(P_0,\delta)$ 分别简记为 $U(P_0)$ 和 $\overset{\circ}{U}(P_0)$.

内点：如果存在点 P 的某个邻域 $U(P,\delta)$,使得 $U(P,\delta) \subset E$,则称 P 为 E 的**内点**.

例如,在图 2.2 中,P_1 为 E 的内点.

外点：如果存在点 P 的某个邻域 $U(P,\delta)$,使得 $U(P,\delta) \cap E = \varnothing$,则称 P 为 E 的**外点**.

例如,在图 2.2 中,P_2 为 E 的外点.

边界点：如果点 P 的任意邻域内既有 E 中的点也有 E 外的点,则称 P 为 E 的**边界点**.

例如,在图 2.2 中,P_3 为 E 的边界点.

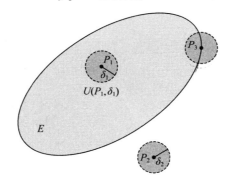

Figure 2.2
图 2.2

Boundary: The set of all boundary points of E is called the **boundary** of E, denoted by ∂E.

Obviously, the interior point of E must belong to E, the exterior point of E must not belong to E, and the boundary point of E may belong to E, or it may not belong to E. There is another relationship between any point P and a point set E, except for the three relationships above, that is the following definition:

Accumulation point: Point P is called the **accumulation point** of E if the deleted neighborhood of P always has a point in E for any $\delta > 0$.

It can be seen from the definition of accumulation point that the accumulation point P of E can belong to E or not belong to E.

According to the characteristics of the points in point set, some important sets of plane points can be defined.

Open set: A point set E is called **open set** if all its points are interior points.

Closed set: A point set E is called **closed set** if its boundary satisfies $\partial E \subset E$.

For example, the point set that satisfies $\{(x,y) \mid 2 < x^2 + y^2 < 3\}$ is an open set; the point set that satisfies $\{(x,y) \mid 2 \leqslant x^2 + y^2 \leqslant 3\}$ is a closed set; the point set that satisfies $\{(x,y) \mid 2 < x^2 + y^2 \leqslant 3\}$ is neither an open set nor a closed set.

Connected set: If any two points in the point set E can be connected by a polyline, and the points on the polyline belong to E, then the point set E is called **connected set**.

Open region: A connected open set is called **open region**.

Closed region: The set of points in the open region along with its boundary is called a **closed region**.

For example, the point set that satisfies $\{(x,y) \mid 2 < x^2 + y^2 < 3\}$ is an open region, but the point set that satisfies $\{(x,y) \mid 2 \leqslant x^2 + y^2 \leqslant 3\}$ is a closed region.

Bounded set: For a planar point set E, if there is a positive number δ that satisfies

边界：E 的边界点的全体称为 E 的边界，记为 ∂E.

显然，E 的内点必属于 E，E 的外点必不属于 E，而 E 的边界点可能属于 E，也可能不属于 E. 任意一点 P 与一个点集 E 之间除了上述三种关系之外，还有另一种关系，就是下面定义的聚点.

聚点：如果对于任意的 $\delta > 0$，点 P 的去心邻域 $\overset{\circ}{U}(P_0, \delta)$ 内总有 E 中的点，则称 P 是 E 的聚点.

由聚点的定义可知，E 的聚点 P 可以属于 E，也可以不属于 E.

根据点集中点的特征，可以定义一些重要的平面点集.

开集：如果点集 E 中的点都是 E 的内点，则称 E 为开集.

闭集：如果点集 E 的边界 $\partial E \subset E$，则称 E 为闭集.

例如，点集 $\{(x,y) \mid 2 < x^2 + y^2 < 3\}$ 是开集；点集 $\{(x,y) \mid 2 \leqslant x^2 + y^2 \leqslant 3\}$ 是闭集；而点集 $\{(x,y) \mid 2 < x^2 + y^2 \leqslant 3\}$ 既非开集，也非闭集.

连通集：如果点集 E 内任何两点，都可以用折线连接起来，且该折线上的点都属于 E，则称 E 为连通集.

开区域：连通的开集称为开区域.

闭区域：开区域连同它的边界一起所构成的点集称为闭区域.

例如，点集 $\{(x,y) \mid 2 < x^2 + y^2 < 3\}$ 是开区域，而点集 $\{(x,y) \mid 2 \leqslant x^2 + y^2 \leqslant 3\}$ 是闭区域.

有界集：对于平面点集 E，如果存在某一正数 δ，使得

$$E \subset U(O,\delta),$$

where O is coordinate origin, then E is called a **bounded set**.

Unbounded set: If a set is not a bounded set, it is called an **unbounded set**.

For example, the point set that satisfies $\{(x,y) | 2 \leqslant x^2 + y^2 \leqslant 3\}$ is a bounded closed set, the point set that satisfies $\{(x,y) | x^2 + y^2 > 1\}$ is an unbounded open region, the point set that satisfies $\{(x,y) | x^2 + y^2 \geqslant 1\}$ is an unbounded closed region.

2. The Concept of the Function of Several Variables

Two kinds of functions have been emphasized so far. They are all functions which the variable y is determined by an argument x, that is the variable y is related to only one variable x and is determined by x. Such functions are called a **function of one variable**. However, in many practical problems, we encounter that a variable is related to and determined by two or more other variables, as shown in the following example.

Example 1 The volume V of cylinder depends on its radius r and its height h. In fact we know that $V = \pi r^2 h$.

When r and h, respectively, take values r_0, h_0, we can get the only value of v_0 correponded with r and h_0 through the above equation. We say that V is a function of r and h, written as

$$V(r,h) = \pi r^2 h \quad (r > 0, h > 0).$$

Example 2 At any given time, the temperature T at a point on the surface of the earth depends on the longitude x and latitude y of the point. We can think of T as a function of the two variables x and y, expressed by $T = f(x,y)$.

Although the background meanings of the above two examples are different, they have the same properties, that is one variable is uniquely determined by the other two variables. The definition of two variable functions can be obtained by abstracting these properties.

Definition 2.1 Let D be a set of points on the plane. According to a corresponding rule f, each point (x,y) in D has a uniquely determined real number z corresponding

$$E \subset U(O,\delta),$$

其中 O 为坐标原点，则称 E 为**有界集**.

无界集：如果一个点集不是有界集，就称这个集合为**无界集**.

例如，点集 $\{(x,y) | 2 \leqslant x^2 + y^2 \leqslant 3\}$ 是有界闭区域，点集 $\{(x,y) | x^2 + y^2 > 1\}$ 是无界开区域，点集 $\{(x,y) | x^2 + y^2 \geqslant 1\}$ 是无界闭区域.

2. 多元函数的概念

至今，我们接触的函数都是因变量 y 由一个自变量 x 确定的函数，即变量 y 只与一个变量 x 有关，且由 x 确定. 这样的函数称为**一元函数**. 但是，在许多实际问题中，我们会遇到一个变量与另外两个变量有关并由它们确定的情形，如下面的例子.

例 1 圆柱的体积 V 取决于它的半径 r 和它的高度 h. 事实上，我们知道 $V = \pi r^2 h$.

当 r,h 取定一组值 r_0, h_0 时，通过上式可以得到唯一的 V_0 值与它们对应. 我们说 V 是 r 和 h 的函数，写作

$$V(r,h) = \pi r^2 h \quad (r > 0, h > 0).$$

例 2 在某一时刻，地球表面某点上的温度 T 取决于该点的经度 x 和纬度 y. 我们可以把 T 看成 x 和 y 的函数，并用 $T = f(x,y)$ 来表示这个函数.

尽管上述两个例子的背景意义不同，但它们具有相同的性质，即一个变量由另外两个变量唯一确定. 通过抽象这一性质可以得到二元函数的定义.

定义 2.1 设 D 为平面上的一个点集. 若按照某对应法则 f，D 中每一点 (x,y) 都有唯一确定的实数 z 与之对应，则称 f 为定义

to it, then f is a **function of two variables** defined on D, denoted as
$$z = f(x, y), \quad (x, y) \in D,$$
where x and y are the **independent variable**, z is **dependent variable**, the set D is the domain of f. We also called z a function of x, y, at the same time say the value of z_0 corresponded by $(x_0, y_0) \in D$ is the **function value** of f at (x_0, y_0), denoted as
$$z_0 = f(x_0, y_0) \quad \text{or} \quad z = f(x, y) \Big|_{\substack{x=x_0 \\ y=y_0}}.$$

The set of all function values is called the **range** of the function, denoted as
$$\{f(x, y) \mid (x, y) \in D\}$$
or
$$\{z \mid z = f(x, y), (x, y) \in D\}.$$

The domain of a function of two variables is a subset of $\mathbf{R}^2$, which is usually a flat area in geometry. Sometimes we also write z which is the function of x, y as
$$z = z(x, y) \quad \text{or} \quad z = \varphi(x, y).$$

Similarly, the concept of the function of two variables can be generalized to the function of three variables $u = f(x, y, z)$ or even the function of n variables
$$u = f(x_1, x_2, \cdots, x_n) \quad (n > 3).$$
Usually the functions of two and more than two variables are called the **functions of several variables**. For the function of several variables $f(x_1, x_2, \cdots, x_n)$, we generally default its domain to be the set of points $(x_1, x_2, \cdots, x_n)$ that make the function of several variables meaningful.

Example 3 Find the domain of the following function and evaluate $f(2, 3)$:

(1) $f(x, y) = \dfrac{\sqrt{x+y-2}}{x-1}$;

(2) $f(x, y) = \dfrac{y \ln(x^2 - y)}{x-1}$.

Solution (1) The expression for $f(x, y)$ makes sense if the denominator is nonzero and the number under the square root is nonnegative, so the domain of $f(x, y)$ is
$$D = \{(x, y) \mid x + y - 2 \geqslant 0, x \neq 1\},$$
where the inequality $x + y - 2 \geqslant 0$, that is $y \geqslant -x + 2$, which describes the points in the domain D lie on or above the line $y = -x + 2$, while $x \neq 1$ means that the points on

在 D 上的**二元函数**，记作
$$z = f(x, y), \quad (x, y) \in D,$$
其中称 x 与 y 为**自变量**，z 为**因变量**，D 为 f 的**定义域**。这时也称 z 为 x, y 的函数，同时称 $(x_0, y_0) \in D$ 所对应的值 z_0 为 f 在点 (x_0, y_0) 处的**函数值**，记作
$$z_0 = f(x_0, y_0) \quad \text{或} \quad z_0 = f(x, y) \Big|_{\substack{x=x_0 \\ y=y_0}}.$$
所有函数值的集合称为该函数的**值域**，记作
$$\{f(x, y) \mid (x, y) \in D\}$$
或
$$\{z \mid z = f(x, y), (x, y) \in D\}.$$

二元函数的定义域是 $\mathbf{R}^2$ 的子集，在几何上它通常是一个平面区域。我们有时也将 z 是 x, y 的函数记为
$$z = z(x, y) \quad \text{或} \quad z = \varphi(x, y).$$

类似地，可以将二元函数的概念推广到三元函数 $u = f(x, y, z)$ 甚至 n 元函数
$$u = f(x_1, x_2, \cdots, x_n) \quad (n > 3).$$
通常将二元及二元以上的函数称为**多元函数**。对于多元函数 $f(x_1, x_2, \cdots, x_n)$，我们一般默认其定义域是使得该多元函数有意义的点 $(x_1, x_2, \cdots, x_n)$ 组成的集合。

例3 求下列函数的定义域和 $f(2, 3)$ 的值：

(1) $f(x, y) = \dfrac{\sqrt{x+y-2}}{x-1}$;

(2) $f(x, y) = \dfrac{y \ln(x^2 - y)}{x-1}$.

解 (1) $f(x, y)$ 的表达式是否有意义取决于分母不为0，同时平方根下的数是非负的，所以 $f(x, y)$ 的定义域为
$$D = \{(x, y) \mid x + y - 2 \geqslant 0, x \neq 1\},$$
其中不等式 $x + y - 2 \geqslant 0$ 即 $y \geqslant -x + 2$，说明定义域 D 中的点位于直线 $y = -x + 2$ 上或其上方；而 $x \neq 1$ 意味着直线 $x = 1$ 上的点不

the line $x=1$ must be excluded from the domain D.

Substitute $x=2, y=3$ into the expression of $f(x,y)$, we get
$$f(2,3)=\frac{\sqrt{2+3-2}}{2-1}=\sqrt{3}.$$

(2) Since $\ln(x^2-y)$ can be defined only when $x^2-y>0$, that is, $x^2>y$, and the dominator is nonzero, so the domain of $f(x,y)$ is
$$D=\{(x,y)\mid x^2>y, x\neq 1\}.$$
This shows that the point in domain D is below the parabola $y=x^2$ and points on line $x=1$ is not in the domain.

Substitute $x=2, y=3$ into the expression of $f(x,y)$, we get
$$f(2,3)=\frac{3\ln(4-3)}{2-1}=3\ln 1=0.$$

Example 4 Find the domain and range of function
$$f(x,y)=\sqrt{16-x^2-y^2}.$$

Solution The domain of $f(x,y)$ is
$$D=\{(x,y)\mid 16-x^2-y^2\geqslant 0\}$$
$$=\{(x,y)\mid x^2+y^2\leqslant 16\}$$
which is the disk with center $(0,0)$ and radius 4.

The range of $f(x,y)$ is
$$\{z\mid z=\sqrt{16-x^2-y^2}, (x,y)\in D\}.$$
Since z is a positive square root, thereby $z\geqslant 0$. Also
$$16-x^2-y^2\leqslant 16,$$
that is
$$\sqrt{16-x^2-y^2}\leqslant 4.$$
So the range of $f(x,y)$ is
$$\{z\mid 0\leqslant z\leqslant 4\}=[0,4].$$

We mean the graph of a function $f(x,y)$ of two variables is the graph of the equation $z=f(x,y)$. This graph will normally be a curved surface. For each point (x,y) in the domain, there exists an unique value z corresponding with it, so each line perpendicular to the Oxy-plane intersects with the surface in at most one point.

在定义域中.

将 $x=2, y=3$ 代入 $f(x,y)$ 的表达式，得
$$f(2,3)=\frac{\sqrt{2+3-2}}{2-1}=\sqrt{3}.$$

（2）因为当 $x^2-y>0$，即 $x^2>y$ 时，$\ln(x^2-y)$ 有意义，又分母不能为 0，故 $f(x,y)$ 的定义域为
$$D=\{(x,y)\mid x^2>y, x\neq 1\}.$$
这表明，定义域 D 中的点在抛物线 $y=x^2$ 的下方，且直线 $x=1$ 上的点不在定义域中.

将 $x=2, y=3$ 代入 $f(x,y)$ 的表达式，得
$$f(2,3)=\frac{3\ln(4-3)}{2-1}=3\ln 1=0.$$

例4 求 $f(x,y)=\sqrt{16-x^2-y^2}$ 的定义域和值域.

解 $f(x,y)$ 的定义域是
$$D=\{(x,y)\mid 16-x^2-y^2\geqslant 0\}$$
$$=\{(x,y)\mid x^2+y^2\leqslant 16\}.$$
这是以坐标原点 $(0,0)$ 为中心，4 为半径的圆及其内部.

$f(x,y)$ 的值域是
$$\{z\mid z=\sqrt{16-x^2-y^2}, (x,y)\in D\}.$$
z 是算数平方根，从而 $z\geqslant 0$，又有
$$16-x^2-y^2\leqslant 16,$$
即
$$\sqrt{16-x^2-y^2}\leqslant 4,$$
所以 $f(x,y)$ 的值域是
$$\{z\mid 0\leqslant z\leqslant 4\}=[0,4].$$

我们说二元函数 $f(x,y)$ 的图形，也就是指方程 $z=f(x,y)$ 的图形. 这个图形通常会是一张曲面. 因为对于定义域中的每个点 (x,y)，都有唯一的值 z 与之对应，所以每条与 Oxy 平面垂直的直线最多与这个曲面相交于一点.

Example 5 Sketch the graph of
$$f(x,y)=\frac{1}{3}\sqrt{36-9x^2-4y^2}.$$

Solution Let $z=\frac{1}{3}\sqrt{36-9x^2-4y^2}$ and note that $z\geqslant 0$. If we square both sides of the equation and simplify them, we obtain the equation
$$9x^2+4y^2+9z^2=36,$$
that we know it is an ellipsoidal equation. The graph of the given function is the upper half of this ellipsoid, it is shown in Figure 2.3.

例 5 画出函数
$$f(x,y)=\frac{1}{3}\sqrt{36-9x^2-4y^2}$$
的图形.

解 令 $z=\frac{1}{3}\sqrt{36-9x^2-4y^2}$,注意 $z\geqslant 0$. 如果我们对两边平方并化简,可以得到方程
$$9x^2+4y^2+9z^2=36.$$
我们知道这是椭球面的方程,所以这个函数的图形就是这个椭球面的上半部分,如图 2.3 所示.

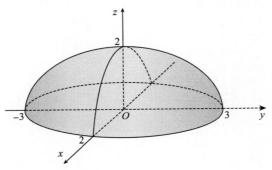

Figure 2.3
图 2.3

2.2 Limit and Continuity of the Function of Two Variables
2.2 多元函数的极限与连续性

This section uses the function of two variables as an example to introduce the limit and continuity of the function of several variables.

本节以二元函数为例来介绍多元函数的极限与连续性.

1. Limit of the Function of Two Variables

Similar to the limit of the function of one variable, the limit of the function of two variables has the general intuitive meaning: the values of $f(x,y)$ get closer and closer to the constant L as (x,y) approaches (x_0,y_0). The problem is that (x,y) can approach (x_0,y_0) in infinite ways.

1. 二元函数的极限

与一元函数的极限类似,二元函数的极限也具有一般的直观意义:当 (x,y) 趋近点 (x_0,y_0) 时,函数 $f(x,y)$ 的值越来越接近某个常数 L. 问题是 (x,y) 可以有无穷多种方式趋近 (x_0,y_0),所以对于二元函数极限的定

So, the definition of the limit of the function of two variables must satisfy that the same constant L can be obtained when (x,y) approach (x_0,y_0) in any way.

Definition 2.2 Let $f(x,y)$ be a function of two variables whose domain D includes points arbitrarily closed to (x_0,y_0). If for any $\varepsilon>0$, there exits a corresponding number $\delta>0$ such that if $(x,y)\in D$ and
$$0<\sqrt{(x-x_0)^2+(y-y_0)^2}<\delta,$$
then
$$|f(x,y)-L|<\varepsilon.$$
Then we say that the **limit** of $f(x,y)$ is L when (x,y) approaches (x_0,y_0) and we write
$$\lim_{(x,y)\to(x_0,y_0)}f(x,y)=L,$$
Other expressions of the limit in the definition are
$$\lim_{\substack{x\to x_0\\y\to y_0}}f(x,y)=L,$$
or
$$f(x,y)\to L,\quad (x,y)\to(x_0,y_0).$$

Notice that $|f(x,y)-L|$ is the distance between $f(x,y)$ and L, and $\sqrt{(x-x_0)^2+(y-y_0)^2}$ is a distance between the point (x,y) and the point (x_0,y_0). Thus Definition 2.2 means that the distance between $f(x,y)$ and L can be arbitrarily small by making the distance from (x,y) to (x_0,y_0) sufficiently small (but nonzero). Figure 2.4 illustrates Definition 2.2 by means of an arrow diagram. For any small interval $(L-\varepsilon,L+\varepsilon)$ containing L, we can find a neighborhood $U(P_0,\delta)$ of point $P_0(x_0,y_0)$ that all the point in $U(P_0,\delta)$ (except for possibly (x_0,y_0)) can be mapped into the interval $(L-\varepsilon,L+\varepsilon)$.

义,要求以任何路径趋近(x_0,y_0)时都得到相同的常数L.

定义2.2 设$f(x,y)$是定义在D上的二元函数,D中包含任意接近(x_0,y_0)的点. 如果对任意的$\varepsilon>0$,存在相应的正数δ,使得对满足
$$0<\sqrt{(x-x_0)^2+(y-y_0)^2}<\delta$$
的$(x,y)\in D$,都有
$$|f(x,y)-L|<\varepsilon.$$
那么我们称(x,y)趋近(x_0,y_0)时,$f(x,y)$的**极限**是L,记作
$$\lim_{(x,y)\to(x_0,y_0)}f(x,y)=L.$$
这个极限的其他表示形式是:
$$\lim_{\substack{x\to x_0\\y\to y_0}}f(x,y)=L,$$
或者
$$f(x,y)\to L,\quad (x,y)\to(x_0,y_0).$$

注意$|f(x,y)-L|$是$f(x,y)$和L之间的距离,$\sqrt{(x-x_0)^2+(y-y_0)^2}$是点$(x,y)$到点$(x_0,y_0)$之间的距离.因此,定义2.2说的是,通过使$(x,y)$到$(x_0,y_0)$的距离足够小(但不为0),可以让$f(x,y)$和$L$之间的距离任意小.图2.4通过箭头的方式说明了定义2.2:对任意包含L的小区间$(L-\varepsilon,L+\varepsilon)$,我们可以找到点$P_0(x_0,y_0)$的一个邻域$U(P_0,\delta)$,把$U(P_0,\delta)$中的所有点(可以不含$(x_0,y_0)$)都映射到区间$(L-\varepsilon,L+\varepsilon)$内.

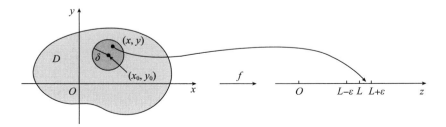

Figure 2.4
图 2.4

Another illustration of Definition 2.2 is given in Figure 2.5 where the surface S is the graph of $f(x,y)$: if $\varepsilon > 0$, we can find $\delta > 0$ such that if (x,y) is in the neighborhood $U(P_0, \delta)$ and $(x,y) \neq (a,b)$, then the corresponding part of S lie between the horizontal planes $z = L - \varepsilon$ and $z = L + \varepsilon$.

定义 2.2 的另一个解释由图 2.5 给出，其中的曲面 S 是函数 $f(x,y)$ 的图形：如果给定 $\varepsilon > 0$，我们可以找到 $\delta > 0$，使得对于被限制在邻域 $U(P_0, \delta)$ 内的点 (x,y)（这里 $(x,y) \neq (a,b)$），曲面 S 的相应部分位于水平面 $z = L - \varepsilon$ 和 $z = L + \varepsilon$ 之间。

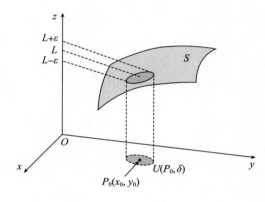

Figure 2.5
图 2.5

It is not hard to deduce from the definition of limit of the function of two variables that local boundedness, feature of guarantee code, sandwich theorem and four fundamental rules of the function of one variable also apply to the limit of the function of two variables.

For functions of one variable, when we let x approach x_0, there are only two possibilities, from the left or from the right tend to x_0. If
$$\lim_{x \to x_0^-} f(x) \neq \lim_{x \to x_0^+} f(x),$$
then $\lim\limits_{x \to a} f(x)$ does not exist.

The function of two variable is not so simple, because we can let (x,y) approach (x_0, y_0) in any way and infinite direction (Figure 2.6) as long as (x,y) stay within the domain D of $f(x,y)$.

If the limit of $f(x,y)$ exists when (x,y) tends to (x_0, y_0), then $f(x,y)$ must approach the same limit value no matter how (x,y) approaches (x_0, y_0). Thus if the function $f(x,y)$ has different limits when (x,y) tends to (x_0, y_0)

由二元函数极限的定义不难推知，一元函数极限的局部有界性、保号性、夹逼准则等性质以及四则运算法则对于二元函数的极限也成立．

对于一元函数来说，当我们令 x 趋近 x_0 时，只有两种可能，从 x_0 的左侧或者右侧趋近 x_0. 如果
$$\lim_{x \to x_0^-} f(x) \neq \lim_{x \to x_0^+} f(x),$$
那么 $\lim\limits_{x \to x_0} f(x)$ 就不存在了．

二元函数就不那么简单了，因为在 $f(x,y)$ 的定义域 D 中，我们可以让 (x,y) 以任意方式、沿无穷个方向趋近 (x_0, y_0)（图 2.6）．

如果 $f(x,y)$ 当 (x,y) 趋近 (x_0, y_0) 时极限存在，指的是无论 (x,y) 以何种方式趋近 (x_0, y_0)，都要求 $f(x,y)$ 必须趋近同一个极限值．因此，如果 (x,y) 按两种不同的路径

along two different paths, or the limit corresponding to one of the paths does not exist, then it follows that $\lim\limits_{(x,y)\to(x_0,y_0)} f(x,y)$ does not exist.

趋近(x_0,y_0)时,函数$f(x,y)$有不同的极限值,或者其中一种路径对应的极限不存在,那么极限$\lim\limits_{(x,y)\to(x_0,y_0)} f(x,y)$便不存在.

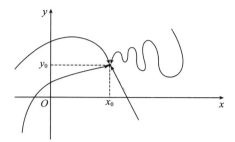

Figure 2.6

图 2.6

Example 1 Prove that the function
$$f(x,y)=\frac{x^2-y^2}{x^2+y^2}$$
has no limit at the origin $(0,0)$.

Solution The function $f(x,y)$ is defined at any point of the Oxy-plane except at the origin $(0,0)$. At all points on the x-axis different from the origin, the value of f is
$$f(x,0)=\frac{x^2-0}{x^2+0}=1.$$
Thus the limit of $f(x,y)$ when (x,y) approaches $(0,0)$ along the x-axis is
$$\lim_{(x,0)\to(0,0)}f(x,0)=\lim_{(x,0)\to(0,0)}\frac{x^2-0}{x^2+0}=1.$$
Similarly, the limit of $f(x,y)$ when (x,y) approaches $(0,0)$ along the y-axis is
$$\lim_{(0,y)\to(0,0)}f(0,y)=\lim_{(0,y)\to(0,0)}\frac{0-y^2}{0+y^2}=-1.$$
Thus, we get different limit values when (x,y) approaches $(0,0)$ in two different paths along x-axis and y-axis. Therefore, the limit of $f(x,y)$ doesn't exist at origin $(0,0)$.

Changing the limit of a function of two variables into a polar coordinate form will be easier to analyze, especially the limits at the origin. In fact, $(x,y)\to(0,0)$ if and only

例1 证明:函数
$$f(x,y)=\frac{x^2-y^2}{x^2+y^2}$$
在原点$(0,0)$处没有极限.

解 除了原点$(0,0)$外,函数$f(x,y)$在Oxy平面上的任意点处都有定义.在x轴上异于原点$(0,0)$的所有点处,$f(x,y)$的值是
$$f(x,0)=\frac{x^2-0}{x^2+0}=1.$$
因此,当(x,y)沿x轴趋于$(0,0)$时,$f(x,y)$的极限为
$$\lim_{(x,0)\to(0,0)}f(x,0)=\lim_{(x,0)\to(0,0)}\frac{x^2-0}{x^2+0}=1.$$
同样,当(x,y)沿y轴趋于$(0,0)$时,$f(x,y)$的极限为
$$\lim_{(0,y)\to(0,0)}f(0,y)=\lim_{(0,y)\to(0,0)}\frac{0-y^2}{0+y^2}=-1.$$
因此,当(x,y)沿x轴和y轴两个不同路径趋于$(0,0)$时,我们得到不同的极限值.因此,函数$f(x,y)$在原点$(0,0)$处的极限不存在.

有时将二元函数的极限变成极坐标的形式来分析是更容易的,尤其是在原点的极限.事实上,由于$(x,y)\to(0,0)$当且仅当$r=$

if $r = \sqrt{x^2 + y^2} \to 0$. Thus, limits of functions of two variables can sometimes be expressed as limits involving just one variable r. For example, we can use polar coordinate to show that the limit in Example 1 doesn't exist:

$$\lim_{(x,y) \to (0,0)} \frac{x^2 - y^2}{x^2 + y^2} = \lim_{r \to 0} \frac{r^2 \cos^2\theta - r^2 \sin^2\theta}{r^2}$$
$$= \lim_{r \to 0} \cos 2\theta = \cos 2\theta,$$

which takes all values between -1 and 1 in every neighborhood of $(0,0)$. We conclude that the limit does not exist.

Example 2 Evaluate the following limits if they exist:

(1) $\lim\limits_{(x,y) \to (0,0)} \dfrac{\sin(x^2 + y^2)}{3x^2 + 3y^2}$;

(2) $\lim\limits_{(x,y) \to (0,0)} \dfrac{xy}{x^2 + y^2}$.

Solution (1) Changing to polar coordinates and using L'Hospital Rule, we have

$$\lim_{(x,y) \to (0,0)} \frac{\sin(x^2 + y^2)}{3x^2 + 3y^2} = \lim_{r \to 0} \frac{\sin r^2}{3r^2}$$
$$= \frac{1}{3} \lim_{r \to 0} \frac{2r \cos r^2}{2r} = \frac{1}{3}.$$

(2) Again, changing to polar coordinates, we have

$$\lim_{(x,y) \to (0,0)} \frac{xy}{x^2 + y^2} = \lim_{r \to 0} \frac{r \cos\theta \cdot r \sin\theta}{r^2}$$
$$= \cos\theta \sin\theta.$$

Since this limit depends on θ. As a result, tending to origin $(0,0)$ along different paths we will get different limits. Thus, this limit does not exist.

Theorem 2.1 If $f(x,y)$ is a polynomial function, then

$$\lim_{(x,y) \to (x_0, y_0)} f(x,y) = f(x_0, y_0),$$

and if $f(x,y)$ is a rational function, that is

$$f(x,y) = \frac{p(x,y)}{q(x,y)},$$

where p and q are polynomial function and $q(x_0, y_0) \neq 0$, then

$$\lim_{(x,y) \to (x_0, y_0)} f(x,y) = \frac{p(x_0, y_0)}{q(x_0, y_0)}.$$

Example 3 Find the following limits if they exist:

(1) $\lim\limits_{(x,y) \to (1,2)} (x^2 y + 3y)$;

(2) $\lim\limits_{(x,y)\to(0,0)} \dfrac{x^2+y^2+1}{x^2-y^2}$.

Solution (1) The function we require to solve the limit is a polynomial, so we have
$$\lim_{(x,y)\to(1,2)} (x^2 y+3y)=1^2 \cdot 2+3 \cdot 2=8.$$

(2) The function is a rational function, but the limit of the denominator is equal to 0, while the limit of the numerator is 1. According to the relationship between infinity and infinitesimal, we have
$$\lim_{(x,y)\to(0,0)} \dfrac{x^2+y^2+1}{x^2-y^2}=\infty.$$
Thus, this limit does not exist.

2. Continuity of the Function of Two Variables

Evaluating limits of continuous function of one variable is easy. It can be accomplished by direct substitution because a continuous function satisfies
$$\lim_{x\to x_0} f(x)=f(x_0),$$
where $f(x)$ is continuous at point x_0. Continuous functions of two variables also have the direct substitution property.

Definition 2.3 Let the function of two variables $f(x,y)$ be defined on D, $(x_0,y_0)\in D$ and D contains any point tend to (x_0,y_0). If
$$\lim_{(x,y)\to(x_0,y_0)} f(x,y)=f(x_0,y_0),$$
we say the function $f(x,y)$ is **continuous** at $P(x_0,y_0)$. Otherwise, the function $f(x,y)$ is **discontinuous** at point $P(x_0,y_0)$. We say function $f(x,y)$ is **continuous function** on D if $f(x,y)$ is continuous at every point (x,y) in D.

We can say that $f(x,y)$ is continuous at the point $P(x_0,y_0)$, and $f(x,y)$ satisfies all the following conditions:

(1) $f(x,y)$ has a value at $P(x_0,y_0)$;

(2) $f(x,y)$ has a limit at $P(x_0,y_0)$;

(3) The value of $f(x,y)$ at $P(x_0,y_0)$ is equal to the limit there. In summary, we require that
$$\lim_{(x,y)\to(x_0,y_0)} f(x,y)=f(x_0,y_0).$$

In follows that, this is essentially the same requirement

for continuity of a function of one variable. Intuitively, this also means that $f(x,y)$ has on jumps, wild fluctuations, or unbounded behavior at $P(x_0,y_0)$. If the function $z=f(x,y)$ is continuous in the plane region D, which is reflected in the geometry, the graph of the function on D is a continuous curved surface without holes(including point holes).

Theorem 2.1 can be used to illustrate that polynomial functions are continuous for all (x,y) and that rational functions are continuous everywhere except for the point where the denominator is equal to 0. Furthermore, As the limit algorithm of the function of one variable also applies to limit of the function of two variable, so sums, differences, products, and quotients (the denominator is nonzero) of continuous functions of two variables are continuous.

Further, the composite function of continuous function of two variables is also continuous, that is, the following theorem is holds.

Theorem 2.2 (Continuity of the Composite Functions)

If a function $u=g(x,y)$ of two variables is continuous at $P_0(x_0,y_0)$ and a function of one variable $z=f(u)$ is continuous at $u=g(x_0,y_0)$, then the composite function
$$f \circ g = (f \circ g)(x,y) = f(g(x,y))$$
is continuous at $P_0(x_0,y_0)$.

Example 4 Discuss the continuity of the following functions:

(1) $f(x,y) = \dfrac{2x+3y}{y-4x^2}$;

(2) $g(x,y) = \cos(x^3 - 4xy + y^2)$.

Solution (1) $f(x,y)$ is a rational function, so according to Theorem 2.1, it is continuous at every point where the denominator is nonzero. The denominator $y - 4x^2$ is equal to zero along the parabola $y = 4x^2$. Thus, $f(x,y)$ is continuous for all (x,y) except for those lying on the parabola $y = 4x^2$.

(2) $h(x,y) = x^3 - 4xy + y^2$ is a polynomial, and continuous for all (x,y). Also, $z = \cos u$ is continuous for every real number u. According to Theorem 2.2, the composite function

函数连续的要求在本质上是相同的. 直观地说, 这也意味着 $f(x,y)$ 在点 $P_0(x_0,y_0)$ 处不发生跳跃、震荡或无界的情况. 如果函数 $z=f(x,y)$ 在平面区域 D 上连续, 反映在几何上, 则是该函数在 D 上的图形是一块连续无洞(包括"点洞")的曲面.

定理 2.1 能用来说明: 对所有 (x,y), 多项式是连续的; 除去分母为零的点, 有理函数也是处处连续的. 此外, 由于一元函数极限的运算法则对于二元函数极限仍然成立, 所以二元连续函数的和、差、积、商(分母不为 0)都是连续的.

进一步, 还可以得到二元连续函数的复合函数也是连续的, 即有下面的定理成立.

定理 2.2(复合函数的连续性)

若二元函数 $u=g(x,y)$ 在点 $P_0(x_0,y_0)$ 处连续, 一元函数 $z=f(u)$ 在点 $u=g(x_0,y_0)$ 处连续, 则复合函数
$$f \circ g = (f \circ g)(x,y) = f(g(x,y))$$
在 $P_0(x_0,y_0)$ 处是连续的.

例 4 讨论下列函数的连续性:

(1) $f(x,y) = \dfrac{2x+3y}{y-4x^2}$;

(2) $g(x,y) = \cos(x^3 - 4xy + y^2)$.

解 (1) 因为 $f(x,y)$ 是有理函数, 所以由定理 2.1 知它在分母不为零的所有点都是连续的. 分母 $y - 4x^2$ 沿抛物线 $y = 4x^2$ 等于零. 因此, 除了抛物线 $y = 4x^2$ 上的点外, $f(x,y)$ 都是连续的.

(2) $h(x,y) = x^3 - 4xy + y^2$ 是多项式, 在任意点 (x,y) 处都连续. 同样, $z = \cos u$ 对于每个实数 u 都是连续的. 根据定理 2.2, 复合函数

$$g(x,y) = \cos(x^3 - 4xy + y^2)$$

is continuous for all (x,y).

According to the four fundamental rules of continuous function of two variables and the continuity of the composite function, we can be easy to get that elementary function of two variables is continuous in its definite area, here the **elementary function of two variables** is taking such a function which can be obtained by a constant and elementary function of one variable taking x and y as its independents variables after finite four fundamental rules and composite the operation. It is a function that can be expressed by an equation. The definited area is the area included in domain.

Example 5 Find the limit
$$\lim_{(x,y)\to(1,2)}(x^2y^3 - x^3y^2 + 3x + 2y).$$

Solution Since $f(x,y) = (x^2y^3 - x^3y^2 + 3x + 2y)$ is a polynomial, it is continuous everywhere in the plane, so we can find the limit by direct substitution:
$$\lim_{(x,y)\to(1,2)}(x^2y^3 - x^3y^2 + 3x + 2y)$$
$$= 1^2 \cdot 2^3 - 1^3 \cdot 2^2 + 3 \cdot 1 + 2 \cdot 2$$
$$= 11.$$

Example 6 Where is the function
$$f(x,y) = \frac{x^2 - y^2}{x^2 + y^2}$$
continuous?

Solution The function $f(x,y)$ is discontinuous at point $(0,0)$ because it is not defined there. Since $f(x,y)$ is a rational function, it is known from Theorem 2.1 $f(x,y)$ is continuous on its domain
$$D = \{(x,y) \mid (x,y) \neq (0,0)\}.$$

Example 7 Find the limit: $\lim_{(x,y)\to(0,0)} \dfrac{\sqrt{xy+1}-1}{xy}$.

Solution $\lim_{(x,y)\to(0,0)} \dfrac{\sqrt{xy+1}-1}{xy}$
$$= \lim_{(x,y)\to(0,0)} \frac{xy+1-1}{xy(\sqrt{xy+1}+1)}$$
$$= \lim_{(x,y)\to(0,0)} \frac{1}{\sqrt{xy+1}+1} = \frac{1}{2}.$$

We know that continuous functions of one variable on

closed intervals have good properties, such as boundedness, existence of the most value and so on. For the function of two variables, there are similarities.

Property 1 (Boundedness Theorem) Suppose that the function of two variables $f(x,y)$ is continuous on the bounded closed area D, then $f(x,y)$ is bounded on D, that is, there exists a constant $M>0$. when $(x,y)\in D$, we have
$$|f(x,y)|\leqslant M.$$

Property 2 (Maximum and Minimum Theorem)
Suppose that the function of two variables $f(x,y)$ is continuous on the bounded closed area D, then $f(x,y)$ can get the maximum value M and the minimum value m, that is, there exist $(x_1,y_1),(x_2,y_2)\in D$, for any $(x,y)\in D$, we have
$$M=f(x_1,y_1)\geqslant f(x,y),$$
$$m=f(x_2,y_2)\leqslant f(x,y).$$

Property 3 (Intermediate Value Theorem) The function of two variables $f(x,y)$, which is continuous on the bounded closed area D, must take any value between the maximum value M and the minimum value m, that is for any $m\leqslant c\leqslant M$, there exists $(x_0,y_0)\in D$ and
$$f(x_0,y_0)=c.$$

The concepts and conclusions of the limit and continuity of the functions of two variables mentioned in this section can be generalized to the functions of three or more variables.

性质 1（有界性定理） 设二元函数 $f(x,y)$ 在有界闭区域 D 上连续，则 $f(x,y)$ 在 D 上有界，即存在常数 $M>0$，使得当 $(x,y)\in D$ 时，有
$$|f(x,y)|\leqslant M.$$

性质 2（最大值和最小值定理）
设二元函数 $f(x,y)$ 在有界闭区域 D 上连续，则 $f(x,y)$ 在 D 上必能取到最大值 M 和最小值 m，即存在 $(x_1,y_1),(x_2,y_2)\in D$，使得对任意 $(x,y)\in D$，有
$$M=f(x_1,y_1)\geqslant f(x,y),$$
$$m=f(x_2,y_2)\leqslant f(x,y).$$

性质 3（介值定理） 在有界闭区域 D 上连续的二元函数 $f(x,y)$，必可取得介于最大值 M 和最小值 m 之间的任何值，即对于任何 $m\leqslant c\leqslant M$，存在 $(x_0,y_0)\in D$，使得
$$f(x_0,y_0)=c.$$

本节中前面关于二元函数的极限和连续性的概念和结论都可以推广到三元及三元以上函数。

2.3 Partial Derivatives
2.3 偏导数

1. Concept of the Partial Derivatives

Suppose that f is a function of two variables x and y. If y is constant, which for example one variable $y=y_0$, then $f(x,y_0)$ is a function of one variable x. Its derivative at $x=x_0$ is called the **partial derivative** of $f(x,y)$ with respect to x at (x_0,y_0) and is written by $f_x(x_0,y_0)$, that is

1. 偏导数的概念

假设 f 是两个变量 x 和 y 的二元函数。若 y 保持不变，比如说 $y=y_0$，则 $f(x,y_0)$ 是只有一个变量 x 的一元函数。$f(x,y_0)$ 在 $x=x_0$ 处的导数称为二元函数 $f(x,y)$ 关于 x 在点 (x_0,y_0) 处的**偏导数**，记为 $f_x(x_0,y_0)$，即

$$f_x(x_0, y_0) = \lim_{h \to 0} \frac{f(x_0+h, y_0) - f(x_0, y_0)}{h}.$$

Similarly, the **partial derivative** of $f(x,y)$ with respect to y at point (x_0, y_0) is denoted by $f_y(x_0, y_0)$, that is

$$f_y(x_0, y_0) = \lim_{h \to 0} \frac{f(x_0, y_0+h) - f(x_0, y_0)}{h}.$$

If we let the point (x_0, y_0) vary in above two functions, $f_x(x_0, y_0)$ and $f_y(x_0, y_0)$ will become functions of two variables x, y. They are called the **partial derivative function** of $f(x,y)$ for x and y, written as $f_x(x,y)$ and $f_y(x,y)$, that is

$$f_x(x,y) = \lim_{h \to 0} \frac{f(x+h, y) - f(x, y)}{h},$$

$$f_y(x,y) = \lim_{h \to 0} \frac{f(x, y+h) - f(x, y)}{h}.$$

In the case of no causing ambiguity, the partial derivative function is also referred to **partial derivative**. The partial derivative of $z = f(x,y)$ can be written as

$$f_x(x,y) = f_x = \frac{\partial f}{\partial x} = \frac{\partial}{\partial x} f(x,y) = \frac{\partial z}{\partial x};$$

$$f_y(x,y) = f_y = \frac{\partial f}{\partial y} = \frac{\partial}{\partial y} f(x,y) = \frac{\partial z}{\partial y}.$$

Example 1 Let

$$f(x,y) = \begin{cases} \dfrac{xy}{x^2+y^2}, & x^2+y^2 \neq 0, \\ 0, & x^2+y^2 = 0. \end{cases}$$

Find $f_x(0,0)$ and $f_y(0,0)$.

Solution According to the definition of by the partial derivative, we have

$$f_x(0,0) = \lim_{h \to 0} \frac{f(0+h, 0) - f(0,0)}{h}$$

$$= \lim_{h \to 0} \frac{\frac{h \cdot 0}{h^2 + 0^2} - 0}{h} = 0.$$

Similarly, we have

$$f_y(0,0) = 0.$$

2. Rule for Finding Partial Derivatives

By the definition of partial derivatives, partial derivative is essentially the derivative of the function of one variable. So we have the following **rules** for finding partial derivatives:

(1) To find f_x, considering y as a constant and finding derivative of $f(x,y)$ with respect to x;

(2) To find f_y, regard x as a constant and finding derivative of $f(x,y)$ with respect to y.

Example 2 Find $f_x(1,2)$ and $f_y(1,2)$ if
$$f(x,y) = x^2 y + 3y^3.$$

Solution To find $f_x(x,y)$, we consider y as a constant and find derivative with respect to x, we have
$$f_x(x,y) = 2xy + 0 = 2xy.$$
Thus,
$$f_x(1,2) = 2 \cdot 1 \cdot 2 = 4.$$
Similarly, we consider x as a constant and finding derivative of with respect to y, we have
$$f_y(x,y) = x^2 + 9y^2.$$
and so
$$f_y(1,2) = 1^2 + 9 \cdot 2^2 = 37.$$

Example 3 If $z = x^2 \sin(xy^2)$, find $\dfrac{\partial z}{\partial x}$ and $\dfrac{\partial z}{\partial y}$.

Solution
$$\begin{aligned}\frac{\partial z}{\partial x} &= x^2 \frac{\partial}{\partial x}[\sin(xy^2)] + \sin(xy^2)\frac{\partial}{\partial x}(x^2)\\ &= x^2 \cos(xy^2)\frac{\partial}{\partial x}(xy^2)\\ &\quad + \sin(xy^2)\cdot 2x\\ &= x^2 \cos(xy^2)\cdot y^2 + 2x\sin(xy^2)\\ &= x^2 y^2 \cos(xy^2) + 2x\sin(xy^2),\\ \frac{\partial z}{\partial y} &= x^2 \cos(xy^2)\cdot 2xy\\ &= 2x^3 y\cos(xy^2).\end{aligned}$$

If $z(x,y)$ is a function of two variables of x,y determined by equation
$$F(x,y,z) = 0,$$
we can also use the derivation rules similar to the implicit function of one variable to find the partial derivatives of $z(x,y)$ about x and y respectively. We illustrate it by the following example.

Example 4 Find $\dfrac{\partial z}{\partial x}$ and $\dfrac{\partial z}{\partial y}$, if z is an implicit function of x and y determined by the equation
$$x^3 + y^3 + z^3 + 6xyz = 1.$$

Solution To find $\dfrac{\partial z}{\partial x}$, we consider y as a constant and find derivates of both sides of equation with respect to x, that is

$$3x^2 + 3z^2 \frac{\partial z}{\partial x} + 6yz + 6xy \frac{\partial z}{\partial x} = 0,$$

where we regard y as a constant. Solving this equation for $\frac{\partial z}{\partial x}$, we obtain

$$\frac{\partial z}{\partial x} = -\frac{x^2 + 2yz}{z^2 + 2xy}.$$

Similarly, regard x as a constant and derivate on both sides of the equation with respect to y, we have

$$3y^2 + 3z^2 \frac{\partial z}{\partial y} + 6xz + 6xy \frac{\partial z}{\partial y} = 0,$$

that is

$$\frac{\partial z}{\partial y} = -\frac{y^2 + 2xz}{z^2 + 2xy}.$$

Remark In Example 3, when y and x is regarded as the constant, z is a composite function. Therefore, the finding derivative with respect to x or y is

$$\frac{\partial}{\partial x}(z^3) = 3z^2 \frac{\partial z}{\partial x},$$

or

$$\frac{\partial}{\partial y}(z^3) = 3z^2 \frac{\partial z}{\partial y}.$$

3. Geometric Interpretations of Partial Derivative

Let $f(x,y)$ be a continuous function defined in area D, $(x_0, y_0) \in D$. Consider the surface determined by the equation $z = f(x, y)$. As shown in Figure 2.7, let the plane $y = y_0$ this surface intersect in the plane curve C_1. Then the equation of curve C_1 on plane $y = y_0$ is

$$z = f(x, y_0).$$

According to the definition of partial derivative, we know

$$f_x(x_0, y_0) = \frac{\mathrm{d}}{\mathrm{d}x} f(x, y_0) \Big|_{x = x_0}.$$

Thus, by the geometric meaning of the derivative of the function of one variable, the value of $f_x(x_0, y_0)$ is the slope on x-axis of the tangent line $M_0 T_1$ to this curve C_1 at $M_0(x_0, y_0, f(x_0, y_0))$. Similarly, if the plane $x = x_0$ intersects the surface in the plane curve C_2, and $f_y(x_0, y_0)$ is the slope on y-axis of the tangent line $M_0 T_2$ to this curve at M_0.

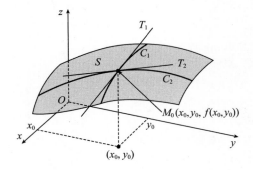

Figure 2.7
图 2.7

Example 5 The surface $z=\sqrt{9-2x^2-y^2}$ and the plane $y=1$ intersect in a curve Γ. Find parametric equations for the tangent line at $(\sqrt{2},1,2)$.

Solution Write $f(x,y)=\sqrt{9-2x^2-y^2}$. Because
$$f_x(x,y)=\frac{1}{2}(9-2x^2-y^2)^{-1/2}(-4x),$$
so
$$f_x(\sqrt{2},1)=-\sqrt{2}.$$
This is the slope on x-axis of the tangent line to the curve Γ at $(\sqrt{2},1,2)$. So on plane $y=1$, the equation for this tangent line is
$$z-2=-\sqrt{2}(x-\sqrt{2}),$$
that is
$$\frac{z-2}{-\sqrt{2}}=\frac{x-\sqrt{2}}{1}.$$
Thus the equation of the tangent line is
$$\begin{cases}\dfrac{z-2}{-\sqrt{2}}=\dfrac{x-\sqrt{2}}{1},\\ y=1.\end{cases}$$
Let
$$\frac{z-2}{-\sqrt{2}}=t,\quad \frac{x-\sqrt{2}}{1}=t,$$
we get the required tangent of the parametric equation of the tangent line
$$\begin{cases}x=\sqrt{2}+t,\\ y=1,\\ z=2-\sqrt{2}t.\end{cases}$$

例5 设曲面 $z=\sqrt{9-2x^2-y^2}$ 和平面 $y=1$ 相交于曲线 Γ,求该曲线在 $(\sqrt{2},1,2)$ 处的切线的参数方程.

解 记 $f(x,y)=\sqrt{9-2x^2-y^2}$. 因为
$$f_x(x,y)=\frac{1}{2}(9-2x^2-y^2)^{-1/2}(-4x),$$
所以
$$f_x(\sqrt{2},1)=-\sqrt{2}.$$
它就是曲线 Γ 在点 $(\sqrt{2},1,2)$ 处的切线关于 x 轴的斜率,所以在平面 $y=1$ 上该切线的方程为
$$z-2=-\sqrt{2}(x-\sqrt{2}),$$
即
$$\frac{z-2}{-\sqrt{2}}=\frac{x-\sqrt{2}}{1}.$$
于是所求的切线方程为
$$\begin{cases}\dfrac{z-2}{-\sqrt{2}}=\dfrac{x-\sqrt{2}}{1},\\ y=1.\end{cases}$$
令
$$\frac{z-2}{-\sqrt{2}}=t,\quad \frac{x-\sqrt{2}}{1}=t,$$
得所求的切线参数方程为
$$\begin{cases}x=\sqrt{2}+t,\\ y=1,\\ z=2-\sqrt{2}t.\end{cases}$$

4. Partial Derivatives of Higher Order

In general, the partial derivatives $f_x(x,y)$, $f_y(x,y)$ of the function of two variables $f(x,y)$ on x and y are still the function of two variable of x, y. Their partial derivatives of x and y are called **second-order partial derivatives** of $f(x,y)$. According to the different order of finding partial derivatives, $f(x,y)$ has four second-order partial derivatives:

$$f_{xx} = \frac{\partial}{\partial x}\left(\frac{\partial f}{\partial x}\right) = \frac{\partial^2 f}{\partial x^2},$$

$$f_{yy} = \frac{\partial}{\partial y}\left(\frac{\partial f}{\partial y}\right) = \frac{\partial^2 f}{\partial y^2},$$

$$f_{xy} = (f_x)_y = \frac{\partial}{\partial y}\left(\frac{\partial f}{\partial x}\right) = \frac{\partial^2 f}{\partial y \partial x},$$

$$f_{yx} = (f_y)_x = \frac{\partial}{\partial x}\left(\frac{\partial f}{\partial y}\right) = \frac{\partial^2 f}{\partial x \partial y},$$

where f_{xy} and f_{yx} are called **second-order mixed partial derivatives** of $f(x,y)$.

Example 6 Find the four second-order partial derivatives of

$$f(x,y) = x e^y - \sin\frac{x}{y} + x^3 y^2.$$

Solution Because

$$f_x(x,y) = e^y - \frac{1}{y}\cos\frac{x}{y} + 3x^2 y^2,$$

$$f_y(x,y) = x e^y + \frac{x}{y^2}\cos\frac{x}{y} + 2x^3 y,$$

so

$$f_{xx}(x,y) = \frac{\partial}{\partial x} f_x(x,y)$$

$$= \frac{1}{y^2}\sin\frac{x}{y} + 6xy^2,$$

$$f_{yy}(x,y) = \frac{\partial}{\partial y} f_y(x,y)$$

$$= x e^y + \frac{x^2}{y^4}\sin\frac{x}{y} - \frac{2x}{y^3}\cos\frac{x}{y} + 2x^3,$$

$$f_{xy}(x,y) = \frac{\partial}{\partial y} f_x(x,y)$$

$$= e^y - \frac{x}{y^3}\sin\frac{x}{y} + \frac{1}{y^2}\cos\frac{x}{y} + 6x^2 y,$$

$$f_{yx}(x,y) = \frac{\partial}{\partial x} f_y(x,y)$$

$$= e^y - \frac{x}{y^3}\sin\frac{x}{y} + \frac{1}{y^2}\cos\frac{x}{y} + 6x^2 y.$$

4. 高阶偏导数

一般而言,二元函数 $f(x,y)$ 关于 x 和关于 y 的偏导数 $f_x(x,y)$ 和 $f_y(x,y)$ 仍然是 x,y 的二元函数,它们关于 x 和关于 y 偏导数称为函数 $f(x,y)$ 的**二阶偏导数**. 根据求偏导数的不同顺序,$f(x,y)$ 有如下四个二阶偏导数:

$$f_{xx} = \frac{\partial}{\partial x}\left(\frac{\partial f}{\partial x}\right) = \frac{\partial^2 f}{\partial x^2},$$

$$f_{yy} = \frac{\partial}{\partial y}\left(\frac{\partial f}{\partial y}\right) = \frac{\partial^2 f}{\partial y^2},$$

$$f_{xy} = (f_x)_y = \frac{\partial}{\partial y}\left(\frac{\partial f}{\partial x}\right) = \frac{\partial^2 f}{\partial y \partial x},$$

$$f_{yx} = (f_y)_x = \frac{\partial}{\partial x}\left(\frac{\partial f}{\partial y}\right) = \frac{\partial^2 f}{\partial x \partial y},$$

其中 f_{xy} 和 f_{yx} 称为 $f(x,y)$ 的**二阶混合偏导数**.

例 6 求函数

$$f(x,y) = x e^y - \sin\frac{x}{y} + x^3 y^2$$

的所有二阶偏导数.

解 因为

$$f_x(x,y) = e^y - \frac{1}{y}\cos\frac{x}{y} + 3x^2 y^2,$$

$$f_y(x,y) = x e^y + \frac{x}{y^2}\cos\frac{x}{y} + 2x^3 y,$$

所以

$$f_{xx}(x,y) = \frac{\partial}{\partial x} f_x(x,y)$$

$$= \frac{1}{y^2}\sin\frac{x}{y} + 6xy^2,$$

$$f_{yy}(x,y) = \frac{\partial}{\partial y} f_y(x,y)$$

$$= x e^y + \frac{x^2}{y^4}\sin\frac{x}{y} - \frac{2x}{y^3}\cos\frac{x}{y} + 2x^3,$$

$$f_{xy}(x,y) = \frac{\partial}{\partial y} f_x(x,y)$$

$$= e^y - \frac{x}{y^3}\sin\frac{x}{y} + \frac{1}{y^2}\cos\frac{x}{y} + 6x^2 y,$$

$$f_{yx}(x,y) = \frac{\partial}{\partial x} f_y(x,y)$$

$$= e^y - \frac{x}{y^3}\sin\frac{x}{y} + \frac{1}{y^2}\cos\frac{x}{y} + 6x^2 y.$$

Notice that in Example 6, $f_{xy} = f_{yx}$. In fact, this is not accidental. When some certain conditions are satisfied, the second-order mixed partial derivative of the function of two variables has nothing to do with the order of finding partial derivative.

Theorem 2.3 If two second-order mixed partial derivative of $f(x,y)$ $f_{xy}(x,y)$ and $f_{yx}(x,y)$ are continuous within area D, then
$$f_{xy}(x,y) = f_{yx}(x,y)$$
at each point within D.

Partial derivatives of the third and higher orders of $f(x,y)$ are defined analogously, and the notation for them is similar. For example for $f(x,y)$, first we find the partial derivative of x, and then find the partial derivative of y twice, we could get the third-order partial derivative f_{xyy}, that is
$$f_{xyy} = \frac{\partial}{\partial y}\left[\frac{\partial}{\partial y}\left(\frac{\partial f}{\partial x}\right)\right] = \frac{\partial}{\partial y}\left(\frac{\partial^2 f}{\partial y \partial x}\right) = \frac{\partial^3 f}{\partial y^2 \partial x}.$$

According to the different order of partial derivatives, there are eight third-order partial derivative of $f(x,y)$, the second-order and above partial derivatives of $f(x,y)$ are collectively referred to the **higher-order partial derivative**.

5. More than Two Variables

The concept of partial derivative of the function of two variables can be generalized to the function of three or more variables. For example, the partial derivative of $f(x,y,z)$ with respect to x at (x,y,z) is denoted by $f_x(x,y,z)$ or $\frac{\partial}{\partial x}f(x,y,z)$ and is written by
$$f_x(x,y,z) = \lim_{h \to 0} \frac{f(x+h,y,z) - f(x,y,z)}{h}.$$

Thus, $f_x(x,y,z)$ can be obtained by considering y and z as constants and differentiating with respect to x.

Similarly, the partial derivatives of $f(x,y,z)$ at point (x,y,z) with respect to y and z, are respectively $f_y(x,y,z)$ and $f_z(x,y,z)$. We also can define that higher-order partial derivatives of $f(x,y,z)$.

Example 7 If the function
$$f(x,y,z) = xy + 2yz + 3zx,$$
find $f_x(x,y,z), f_y(x,y,z), f_z(x,y,z)$.

注意到在例6中，$f_{xy} = f_{yx}$. 其实，这并不是偶然现象，当满足一定条件时，二元函数的二阶混合偏导数与求偏导数的顺序无关.

定理 2.3 如果函数 $f(x,y)$ 的两个二阶混合偏导数 $f_{xy}(x,y)$ 和 $f_{yx}(x,y)$ 在区域 D 内连续，则在 D 内每一点都有
$$f_{xy}(x,y) = f_{yx}(x,y).$$

类似地，可定义函数 $f(x,y)$ 的三阶偏导数和更高阶的偏导数，符号也是相似的. 例如，对于函数 $f(x,y)$，先关于 x 求一阶偏导数，再关于 y 求两次偏导数，可以得到三阶偏导数 f_{xyy}，即
$$f_{xyy} = \frac{\partial}{\partial y}\left[\frac{\partial}{\partial y}\left(\frac{\partial f}{\partial x}\right)\right] = \frac{\partial}{\partial y}\left(\frac{\partial^2 f}{\partial y \partial x}\right) = \frac{\partial^3 f}{\partial y^2 \partial x}.$$
按不同的求偏导数顺序，$f(x,y)$ 共有八个三阶偏导数. 二阶及二阶以上的偏导数统称为**高阶偏导数**.

5. 多于两个变量的情形

二元函数偏导数的概念可推广到三元及三元以上的函数上. 例如，函数 $f(x,y,z)$ 在点 (x,y,z) 处关于 x 的偏导数表示为 $f_x(x,y,z)$ 或 $\frac{\partial}{\partial x}f(x,y,z)$，定义为
$$f_x(x,y,z) = \lim_{h \to 0} \frac{f(x+h,y,z) - f(x,y,z)}{h}.$$

因此，$f_x(x,y,z)$ 可通过把 y 和 z 看作常数，然后对 x 求导数而得到.

类似地，可定义函数 $f(x,y,z)$ 在点 (x,y,z) 处关于 y 和关于 z 的偏导数 $f_y(x,y,z)$ 和 $f_z(x,y,z)$. 我们还可以进一步定义函数 $f(x,y,z)$ 的高阶偏导数.

例 7 若函数
$$f(x,y,z) = xy + 2yz + 3zx,$$
求 $f_x(x,y,z), f_y(x,y,z), f_z(x,y,z)$.

Solution To find $f_x(x,y,z)$, we consider y and z as constants and derive $f(x,y,z)$ with respect to x. Thus,
$$f_x(x,y,z) = y + 3z.$$
To find $f_y(x,y,z)$, we consider x and z as constants and derive $f(x,y,z)$ with respect to y:
$$f_y(x,y,z) = x + 2z.$$
Similarly, we have
$$f_z(x,y,z) = 2y + 3x.$$

Example 8 If the function
$$f(w,x,y,z) = ze^{w^2+x^2+y^2},$$
find all first-order partial derivatives of $f(w,x,y,z)$ and $\dfrac{\partial^2 f}{\partial w \partial x}, \dfrac{\partial^2 f}{\partial x \partial w}, \dfrac{\partial^2 f}{\partial z^2}$.

Solution Four first-order partial derivatives of $f(w,x,y,z)$ are
$$\frac{\partial f}{\partial w} = \frac{\partial}{\partial w}(ze^{w^2+x^2+y^2}) = 2wze^{w^2+x^2+y^2},$$
$$\frac{\partial f}{\partial x} = \frac{\partial}{\partial x}(ze^{w^2+x^2+y^2}) = 2xze^{w^2+x^2+y^2},$$
$$\frac{\partial f}{\partial y} = \frac{\partial}{\partial y}(ze^{w^2+x^2+y^2}) = 2yze^{w^2+x^2+y^2},$$
$$\frac{\partial f}{\partial z} = \frac{\partial}{\partial z}(ze^{w^2+x^2+y^2}) = e^{w^2+x^2+y^2}.$$

The other required second-order partial derivatives are
$$\frac{\partial^2 f}{\partial w \partial x} = \frac{\partial^2}{\partial w \partial x}(ze^{w^2+x^2+y^2})$$
$$= \frac{\partial}{\partial w}(2xze^{w^2+x^2+y^2})$$
$$= 4wxze^{w^2+x^2+y^2},$$
$$\frac{\partial^2 f}{\partial x \partial w} = \frac{\partial^2}{\partial x \partial w}(ze^{w^2+x^2+y^2})$$
$$= \frac{\partial}{\partial x}(2wze^{w^2+x^2+y^2})$$
$$= 4wxze^{w^2+x^2+y^2},$$
$$\frac{\partial^2 f}{\partial z^2} = \frac{\partial^2}{\partial z^2}(ze^{w^2+x^2+y^2})$$
$$= \frac{\partial}{\partial z}(e^{w^2+x^2+y^2}) = 0.$$

解 为了得到 $f_x(x,y,z)$，我们将 y 和 z 视为常数，然后关于 x 求导数. 因此
$$f_x(x,y,z) = y + 3z.$$
为了求得 $f_y(x,y,z)$，我们把 x 和 z 视为常数，然后关于 y 求导数：
$$f_y(x,y,z) = x + 2z.$$
同样，有
$$f_z(x,y,z) = 2y + 3x.$$

例 8 若函数
$$f(w,x,y,z) = ze^{w^2+x^2+y^2},$$
求出 $f(w,x,y,z)$ 的所有一阶偏导数和 $\dfrac{\partial^2 f}{\partial w \partial x}, \dfrac{\partial^2 f}{\partial x \partial w}, \dfrac{\partial^2 f}{\partial z^2}$.

解 $f(w,x,y,z)$ 的四个一阶偏导数为
$$\frac{\partial f}{\partial w} = \frac{\partial}{\partial w}(ze^{w^2+x^2+y^2}) = 2wze^{w^2+x^2+y^2},$$
$$\frac{\partial f}{\partial x} = \frac{\partial}{\partial x}(ze^{w^2+x^2+y^2}) = 2xze^{w^2+x^2+y^2},$$
$$\frac{\partial f}{\partial y} = \frac{\partial}{\partial y}(ze^{w^2+x^2+y^2}) = 2yze^{w^2+x^2+y^2},$$
$$\frac{\partial f}{\partial z} = \frac{\partial}{\partial z}(ze^{w^2+x^2+y^2}) = e^{w^2+x^2+y^2}.$$

其他所求的二阶偏导数为
$$\frac{\partial^2 f}{\partial w \partial x} = \frac{\partial^2}{\partial w \partial x}(ze^{w^2+x^2+y^2})$$
$$= \frac{\partial}{\partial w}(2xze^{w^2+x^2+y^2})$$
$$= 4wxze^{w^2+x^2+y^2},$$
$$\frac{\partial^2 f}{\partial x \partial w} = \frac{\partial^2}{\partial x \partial w}(ze^{w^2+x^2+y^2})$$
$$= \frac{\partial}{\partial x}(2wze^{w^2+x^2+y^2})$$
$$= 4wxze^{w^2+x^2+y^2},$$
$$\frac{\partial^2 f}{\partial z^2} = \frac{\partial^2}{\partial z^2}(ze^{w^2+x^2+y^2})$$
$$= \frac{\partial}{\partial z}(e^{w^2+x^2+y^2}) = 0.$$

2.4 Total Differential
2.4 全微分

1. The Concept of Total Differential

Let us first recall that the differential of the function of one variable $y=f(x)$ at x. If the increment Δy of the function at x can be expressed as
$$\Delta y = A\Delta x + o(\Delta x),$$
where A is related to x, while independent of Δx, $o(\Delta x)$ is the infinitesimal of higher order relative to Δx. Then $A\Delta x$ is the differential of the function of one variable $y=f(x)$ at x, we call the function differentiable at the point x. It can be seen that the concept of differentiation of the function of one variable describes the relationship between increments Δy of function and independent variable Δx. In particular, when $|\Delta x|$ is sufficiently small, the derivative $A\Delta x$ can be used to replace the incremental Δy of the function.

For the function of two variables $z=f(x,y)$, which has two independent variables x and y, when x and y respectively obtain the increment Δx and Δy, the increment of the function is
$$\Delta z = f(x+\Delta x, y+\Delta y) - f(x,y),$$
which is usually called **the total differential** of the function $z=f(x,y)$.

In order to discuss the relationship between the total increment of the function of two variables and the increment of two independent variables, we introduce the definition of the total differential of the function of two variables.

Definition 2.4 Supposed that a function of two variables $z=f(x,y)$ exists in the neighborhood of (x,y) and $(x+\Delta x, y+\Delta y)$ is in this neighborhood, if the total increment of the function at (x,y), which is
$$\Delta z = f(x+\Delta x, y+\Delta y) - f(x,y),$$
can be described as
$$\Delta z = A\Delta x + B\Delta y + o(\rho),$$

1. 全微分的概念

我们先来回忆一下一元函数 $y=f(x)$ 在点 x 处的微分定义:如果函数 $y=f(x)$ 在点 x 处的增量 Δy 可以表示成
$$\Delta y = A\Delta x + o(\Delta x),$$
其中 A 只与点 x 有关而与 Δx 无关,$o(\Delta x)$ 是 Δx 的高阶无穷小量,那么称 $A\Delta x$ 是函数 $f(x)$ 在 x 处的微分,并称函数 $f(x)$ 在点 x 处可微. 可见,一元函数微分的概念描述的是函数增量 Δy 与自变量增量 Δx 的关系. 尤其是,当 $|\Delta x|$ 充分小时,可用微分 $A\Delta x$ 来近似代替函数增量 Δy.

对于二元函数 $z=f(x,y)$,它有两个自变量 x 和 y,当 x 和 y 分别取得增量 Δx,Δy 时,函数的增量为
$$\Delta z = f(x+\Delta x, y+\Delta y) - f(x,y),$$
通常称之为函数 $z=f(x,y)$ 的**全增量**.

为了讨论二元函数的全增量与其两个自变量增量之间的关系,我们引入二元函数全微分的定义.

定义 2.4 设二元函数 $z=f(x,y)$ 在点 (x,y) 的某个邻域内有定义,且点 $(x+\Delta x, y+\Delta y)$ 在该邻域内. 如果函数在点 (x,y) 的全增量
$$\Delta z = f(x+\Delta x, y+\Delta y) - f(x,y)$$
可以表示为
$$\Delta z = A\Delta x + B\Delta y + o(\rho),$$

where A and B has no connection with Δx and Δy, just related to x and y, $\rho=\sqrt{(\Delta x)^2+(\Delta y)^2}$, $o(\rho)$ is a infinitesimal of higher order of ρ as $\rho\to 0$, then the function $f(x,y)$ is **differential** at (x,y), $A\Delta x+B\Delta y$ **the total differential** of function $f(x,y)$ at (x,y), denoted $\mathrm{d}z$ as
$$\mathrm{d}z=A\Delta x+B\Delta y.$$

If the function $z=f(x,y)$ is differentiable at every point in area D, then we call $z=f(x,y)$ **differentiable** in D.

It's similar to the function of one variable. If the function of two variables $z=f(x,y)$ is differentiable at (x,y), then $z=f(x,y)$ is continuous at (x,y). In fact, we have
$$\lim_{\rho\to 0}\Delta z=0.$$
Thus, $z=f(x,y)$ is continuous at point (x,y).

If the function is continuous at (x_0,y_0), how to find A, B? About this question, there is the following theorem.

Theorem 2.4 (The Necessary Condition of Differential Function) If function $z=f(x,y)$ is differentiable at (x,y), then the partial derivative $\frac{\partial z}{\partial x}$, $\frac{\partial z}{\partial y}$ of $f(x,y)$ at (x,y) exit and
$$A=\frac{\partial z}{\partial x},\quad B=\frac{\partial z}{\partial y}.$$

The differential and derivative of the function of one variable are equivalent. But it is invalid in the function of several variables. For example the partial derivative of the function
$$f(x,y)=\begin{cases}\dfrac{xy}{x^2+y^2}, & x^2+y^2\neq 0,\\ 0, & x^2+y^2=0\end{cases}$$
exist (see the example in Section 2.3), but $f(x,y)$ is not continuous at $(0,0)$. It is obvious that $f(x,y)$ is not differential at $(0,0)$.

Theorem 2.5 (The Sufficient Condition of Differentiable)

If the partial derivative of function $f(x,y)$ is continuous at (x,y), $f(x,y)$ is differential at this point.

The concept of total differential can be generalized to function of three or more variables, for example, if the function of three variables $u=f(x,y,z)$ has a continuous partial derivative, then
$$\mathrm{d}u=\frac{\partial u}{\partial x}\mathrm{d}x+\frac{\partial u}{\partial y}\mathrm{d}y+\frac{\partial u}{\partial z}\mathrm{d}z.$$

其中 A,B 与 $\Delta x,\Delta y$ 无关，只与 x,y 有关，$\rho=\sqrt{(\Delta x)^2+(\Delta y)^2}$，$o(\rho)$ 是当 $\rho\to 0$ 时 ρ 的高阶无穷小量，则称函数 $f(x,y)$ 在点 (x,y) 处**可微**，并称 $A\Delta x+B\Delta y$ 是函数 $f(x,y)$ 在点 (x,y) 处的**全微分**，记作 $\mathrm{d}z$，即
$$\mathrm{d}z=A\Delta x+B\Delta y.$$

如果函数 $z=f(x,y)$ 在区域 D 内每一点都可微，则称函数 $f(x,y)$ 在区域 D 内**可微**.

与一元函数类似，如果二元函数 $z=f(x,y)$ 在点 (x,y) 处可微，则 $z=f(x,y)$ 在点 (x,y) 处一定连续. 事实上，由可微可得
$$\lim_{\rho\to 0}\Delta z=0,$$
从而 $z=f(x,y)$ 在点 (x,y) 处连续.

如果函数在点 (x,y) 处可微，如何求 A,B 呢？对于这一问题，有如下定理：

定理 2.4（可微的必要条件） 如果函数 $z=f(x,y)$ 在点 (x,y) 处可微，则函数 $f(x,y)$ 在点 (x,y) 处的偏导数 $\frac{\partial z}{\partial x}$，$\frac{\partial z}{\partial y}$ 存在，并且
$$A=\frac{\partial z}{\partial x},\quad B=\frac{\partial z}{\partial y}.$$

一元函数中，可微与可导是等价的，但在多元函数中，这个结论并不成立. 例如，函数
$$f(x,y)=\begin{cases}\dfrac{xy}{x^2+y^2}, & x^2+y^2\neq 0,\\ 0, & x^2+y^2=0\end{cases}$$
在点 $(0,0)$ 处的两个偏导数存在（见 2.3 节的例 1），但是不难证明 $f(x,y)$ 在点 $(0,0)$ 处不连续，从而 $f(x,y)$ 在点 $(0,0)$ 处不可微.

定理 2.5（可微的充分条件）

若函数 $f(x,y)$ 在点 (x,y) 处的两个偏导数连续，则函数 $f(x,y)$ 在该点一定可微.

全微分的概念可以推广到三元或三元以上的函数. 例如，若三元函数 $u=f(x,y,z)$ 具有连续偏导数，则其全微分的表达式为
$$\mathrm{d}u=\frac{\partial u}{\partial x}\mathrm{d}x+\frac{\partial u}{\partial y}\mathrm{d}y+\frac{\partial u}{\partial z}\mathrm{d}z.$$

Example 1　Find the full increment and total differential of $z = x^2y^2$ at $(2, -1)$, when $\Delta x = 0.02$, $\Delta y = -0.01$.

Solution　The required full increment
$$\Delta z = f(x_0 + \Delta x, y_0 + \Delta y) - f(x_0, y_0)$$
$$= (2+0.02)^2(-1-0.01)^2 - 2^2 \times (-1)^2$$
$$= 0.1624.$$

Two partial derivatives of function $z = x^2y^2$,
$$\frac{\partial z}{\partial x} = 2xy^2, \quad \frac{\partial z}{\partial y} = 2x^2y$$
are continuous, so the total differential exits, also
$$\left.\frac{\partial z}{\partial x}\right|_{(2,-1)} = 2 \times 2 \times (-1)^2 = 4,$$
$$\left.\frac{\partial z}{\partial y}\right|_{(2,-1)} = 2 \times 2^2 \times (-1) = -8,$$
$$dx = \Delta x = 0.02,$$
$$dy = \Delta y = -0.01.$$
then the total differential of $z = x^2y^2$ at $(2, -1)$ is
$$dz = \left.\frac{\partial z}{\partial x}\right|_{(2,-1)} dx + \left.\frac{\partial z}{\partial y}\right|_{(2,-1)} dy$$
$$= 4 \times 0.02 + (-8) \times (-0.01)$$
$$= 0.16.$$

Example 2　Find the total differential of
$$z = e^x \sin(x+y).$$

Solution　Since
$$\frac{\partial z}{\partial x} = e^x \sin(x+y) + e^x \cos(x+y),$$
$$\frac{\partial z}{\partial y} = e^x \cos(x+y),$$
so
$$dz = \frac{\partial z}{\partial x} dx + \frac{\partial z}{\partial y} dy$$
$$= e^x[\sin(x+y) + \cos(x+y)]dx$$
$$+ e^x \cos(x+y) dy.$$

2. The Application of Total Differential in Approximate Calculation

According to the definition of the function of two variables and the sufficient conditions of derivable, if the function $f(x,y)$ have continuous partial derivatives $f_x(x,y)$ and $f_y(x,y)$ in a certain neighborhood $U(P_0, \delta)$ of the point $P_0(x_0, y_0)$, there is an approximate formula

例1　求函数 $z = x^2y^2$ 在点 $(2, -1)$ 处,当 $\Delta x = 0.02$, $\Delta y = -0.01$ 时的全增量与全微分.

解　由定义知,所求的全增量为
$$\Delta z = f(x_0 + \Delta x, y_0 + \Delta y) - f(x_0, y_0)$$
$$= (2+0.02)^2(-1-0.01)^2 - 2^2 \times (-1)^2$$
$$= 0.1624.$$

函数 $z = x^2y^2$ 的两个偏导数为
$$\frac{\partial z}{\partial x} = 2xy^2, \quad \frac{\partial z}{\partial y} = 2x^2y,$$
它们都是连续的,所以全微分是存在的,又
$$\left.\frac{\partial z}{\partial x}\right|_{(2,-1)} = 2 \times 2 \times (-1)^2 = 4,$$
$$\left.\frac{\partial z}{\partial y}\right|_{(2,-1)} = 2 \times 2^2 \times (-1) = -8,$$
$$dx = \Delta x = 0.02,$$
$$dy = \Delta y = -0.01.$$
于是在点 $(2, -1)$ 处的全微分为
$$dz = \left.\frac{\partial z}{\partial x}\right|_{(2,-1)} dx + \left.\frac{\partial z}{\partial y}\right|_{(2,-1)} dy$$
$$= 4 \times 0.02 + (-8) \times (-0.01)$$
$$= 0.16.$$

例2　求函数 $z = e^x \sin(x+y)$ 的全微分.

解　因为
$$\frac{\partial z}{\partial x} = e^x \sin(x+y) + e^x \cos(x+y),$$
$$\frac{\partial z}{\partial y} = e^x \cos(x+y),$$
所以
$$dz = \frac{\partial z}{\partial x} dx + \frac{\partial z}{\partial y} dy$$
$$= e^x[\sin(x+y) + \cos(x+y)]dx$$
$$+ e^x \cos(x+y) dy.$$

2. 全微分在近似计算中的应用

根据二元函数全微分的定义和可微的充分条件,若函数 $f(x,y)$ 在点 $P_0(x_0, y_0)$ 的某个邻域 $U(P_0, \delta)$ 中具有连续的偏导数 $f_x(x,y)$, $f_y(x,y)$,则当 $|\Delta x| = |x - x_0|$, $|\Delta y| = |y - y_0|$ 充分小时,有近似计算公式

$\Delta z \approx \mathrm{d}z$
$$= f_x(x_0, y_0)\Delta x + f_y(x_0, y_0)\Delta y, \quad (2.1)$$
when $|\Delta x| = |x - x_0|$ and $|\Delta y| = |y - y_0|$ are sufficiently small, where the error is a high-order infinitesmal of $\rho = \sqrt{(\Delta x)^2 + (\Delta y)^2}$. The total increment is
$$\Delta z = f(x_0 + \Delta x, y_0 + \Delta y) - f(x_0, y_0).$$
The approximate calculation formula of the function value can be obtained by equation (2.1):
$$f(x_0 + \Delta x, y_0 + \Delta y)$$
$$\approx f(x_0, y_0) + f_x(x_0, y_0)\Delta x$$
$$+ f_y(x_0, y_0)\Delta y. \quad (2.2)$$

Example 3 Find the approximation of $(1.03)^{2.02}$.

Solution Let the function
$$f(x, y) = x^y,$$
then $(1.03)^{2.03}$ is the value of the function $f(1.03, 2.02)$. Due to
$$f(1.03, 2.02) = f(1 + 0.03, 2 + 0.02),$$
if we take $x = 1, \Delta x = 0.03, y = 2, \Delta y = 0.02$, by equation (2.2) there is
$$(1.03)^{2.02} = f(1 + 0.03, 2 + 0.02)$$
$$\approx f(1, 2) + f_x(1, 2) \times 0.03$$
$$+ f_y(1, 2) \times 0.02.$$
And because
$$f_x(x, y) = y x^{y-1},$$
$$f_y(x, y) = x^y \ln x,$$
we have
$$f_x(1, 2) = 2, \quad f_y(1, 2) = 0,$$
so
$$(1.03)^{2.2} \approx 1^2 + 2 \times 0.03 + 0 \times 0.02$$
$$= 1.06.$$

2.5 The Derivative Rule of Multivariate Composite Function

Now the Chain Rule for derivative rule of composite functions of one variable is familiar: if $y = f(x(t))$, where both $f(x)$ and $x(t)$ are differentiable, then

$$\frac{dy}{dt} = \frac{dy}{dx} \cdot \frac{dx}{dt}.$$

In this section, our goal is to generalize this principle to the function of several variables.

Theorem 2.6 (Chain Rule) Let $x = x(t)$ and $y = y(t)$ be derivable at t, and let the function $z = (x, y)$ be differentiable at $(x(t), y(t))$, then the composite function $z = f(x(t), y(t))$ is derivable at t and

$$\frac{dz}{dt} = \frac{\partial z}{\partial x} \cdot \frac{dx}{dt} + \frac{\partial z}{\partial y} \cdot \frac{dy}{dt}. \quad (2.3)$$

Example 1 Suppose that $z = x^3 y$, where $x = 2t$ and $y = t^2$, find $\frac{dz}{dt}$.

Solution $z = x^3 y$ is a composite function. By equation (2.3) we have

$$\begin{aligned}\frac{dz}{dt} &= \frac{\partial z}{\partial x} \cdot \frac{dx}{dt} + \frac{\partial z}{\partial y} \cdot \frac{dy}{dt} \\ &= 3x^2 y \cdot 2 + x^3 \cdot 2t \\ &= 6 \cdot (2t)^2 \cdot t^2 + 2(2t)^3 \cdot t \\ &= 40 t^4.\end{aligned}$$

We can also solve the Example 1 with using Chain Rule. By direct substitution

$$z = x^3 y = (2t)^3 t^2 = 8 t^5,$$

so $\frac{dz}{dt} = 40 t^4$. However, sometimes the direct substitution method is unavailable or inconvenient.

Example 2 As a cylinder is heated, its radius r and height h increase, hence, so does its surface area S. Suppose that at the instant $r = 10$ cm and $h = 100$ cm, r is increasing with 0.2 cm/h and h is increasing with 0.5 cm/h. What is the growth rate of S at this instant?

Solution The formula for total surface area of a cylinder is

$$S = 2\pi r h + 2\pi r^2,$$

and r and h are the functions of t. Thus,

$$\begin{aligned}\frac{dS}{dt} &= \frac{\partial S}{\partial r} \cdot \frac{dr}{dt} + \frac{\partial S}{\partial h} \cdot \frac{dh}{dt} \\ &= (2\pi h + 4\pi r)\frac{dr}{dt} + 2\pi r \frac{dh}{dt}.\end{aligned}$$

When $r = 10$ cm and $h = 100$ cm,

$$\frac{dr}{dt} = 0.2 \text{ cm/h},$$

2.5 The Derivative Rule of Multivariate Composite Function

$$\frac{dh}{dt} = 0.5 \text{ cm/h}.$$

So the growth speed of S is

$$\frac{dS}{dt} = (2\pi \times 100 + 4\pi \times 10) \times 0.2$$
$$+ 2\pi \times 10 \times 0.5$$
$$= 58\pi \text{ (unit: cm}^2/\text{h)}.$$

The result of Theorem 2.6 can be generalized to the compound function of more than two mediate variables. For example, set $w = f(x, y, z)$ be differentiable, and $x = x(t), y = y(t), z = z(t)$ be derivable, then

$$\frac{dw}{dt} = \frac{\partial w}{\partial x} \cdot \frac{dx}{dt} + \frac{\partial w}{\partial y} \cdot \frac{dy}{dt} + \frac{\partial w}{\partial z} \cdot \frac{dz}{dt}. \quad (2.4)$$

Example 3 Suppose that the function
$$w = x^2 y + y + xz,$$
where $x = \cos\theta$, $y = \sin\theta$, and $z = \theta^2$. Find $\dfrac{dw}{d\theta}$ and calculate its value at $\theta = \dfrac{\pi}{3}$.

Solution By the formula (2.4), there is
$$\frac{dw}{d\theta} = \frac{\partial w}{\partial x} \cdot \frac{dx}{d\theta} + \frac{\partial w}{\partial y} \cdot \frac{dy}{d\theta} + \frac{\partial w}{\partial z} \cdot \frac{dz}{d\theta}$$
$$= (2xy + z)(-\sin\theta) + (x^2 + 1)\cos\theta + x \cdot 2\theta$$
$$= -2\cos\theta \sin^2\theta - \theta^2 \sin\theta + \cos^3\theta + \cos\theta$$
$$+ 2\theta\cos\theta$$

At $\theta = \dfrac{\pi}{3}$,
$$\frac{dw}{d\theta} = -2\cos\frac{\pi}{3}\left(\sin\frac{\pi}{3}\right)^2 - \left(\frac{\pi}{3}\right)^2 \sin\frac{\pi}{3}$$
$$+ \left(\cos\frac{\pi}{3}\right)^3 + \cos\frac{\pi}{3} + 2 \cdot \frac{\pi}{3}\cos\frac{\pi}{3}$$
$$= -2 \cdot \frac{1}{2} \cdot \frac{3}{4} - \frac{\pi^2}{9} \cdot \frac{\sqrt{3}}{2} + \frac{1}{8}$$
$$+ \frac{1}{2} + \frac{2\pi}{3} \cdot \frac{1}{2}$$
$$= -\frac{1}{8} - \frac{\pi^2 \sqrt{3}}{18} + \frac{\pi}{3}.$$

The front is derivation rules of the composition between the function of one variable and the function of two variables. Next we consider the case of the composition between two functions of several variables.

Theorem 2.7 (Chain Rule) Let $x = x(s, t)$ and $y = y(s, t)$ have first partial derivatives at (s, t), and let $z = $

$f(x,y)$ be differentiable at $(x(s,t),y(s,t))$. Then composite function $z=f(x(s,t),y(x,t))$ has first partial derivatives at point (s,t) given by

$$\frac{\partial z}{\partial s}=\frac{\partial z}{\partial x}\cdot\frac{\partial x}{\partial s}+\frac{\partial z}{\partial y}\cdot\frac{\partial y}{\partial s},$$

$$\frac{\partial z}{\partial t}=\frac{\partial z}{\partial x}\cdot\frac{\partial x}{\partial t}+\frac{\partial z}{\partial y}\cdot\frac{\partial y}{\partial t}.$$

Example 4 If
$$z=3x^2-y^2,$$
where $x=2s+7t$ and $y=5st$. Find $\dfrac{\partial z}{\partial t}$ and express it with s and t.

Solution
$$\frac{\partial z}{\partial t}=\frac{\partial z}{\partial x}\cdot\frac{\partial x}{\partial t}+\frac{\partial z}{\partial y}\cdot\frac{\partial y}{\partial t}$$
$$=6x\cdot 7+(-2y)\cdot 5s$$
$$=42(2s+7t)-10st\cdot 5s$$
$$=84s+294t-50s^2t.$$

Of course, if we substitute the expressions of x and y into the expression of z and then find the partial derivative with respect to t, we get the same answer:

$$\frac{\partial z}{\partial t}=\frac{\partial}{\partial t}[3(2s+7t)^2-(5st)^2]$$
$$=\frac{\partial}{\partial t}(12s^2+84st+147t^2-25s^2t^2)$$
$$=84s+294t-50s^2t.$$

Similarly, the result of Theorem 2.5 can also be generalized to more than two intermediate variables.

Example 5 If
$$w=x^2+y^2+z^2+xy,$$
where $x=st, y=s-t$, and $z=s+2t$, find $\dfrac{\partial w}{\partial t}$.

Solution There are three intermediate variables x, y, z, so
$$\frac{\partial w}{\partial t}=\frac{\partial w}{\partial x}\cdot\frac{\partial x}{\partial t}+\frac{\partial w}{\partial y}\cdot\frac{\partial y}{\partial t}+\frac{\partial w}{\partial z}\cdot\frac{\partial z}{\partial t}$$
$$=(2st+s-t)s+(2s-2t+st)\cdot(-1)$$
$$+(2s+4t)\cdot 2$$
$$=2s^2t+s^2-2st+2s+10t.$$

Example 6 If
$$u=x^4y+y^2z^3,$$
where $x=rse^t, y=rs^2e^{-t}$, and $z=r^2s\sin t$, find the value of $\dfrac{\partial u}{\partial s}$ when $r=2, s=1, t=0$.

Solution We have
$$\frac{\partial u}{\partial s}=\frac{\partial u}{\partial x}\cdot\frac{\partial x}{\partial s}+\frac{\partial u}{\partial y}\cdot\frac{\partial y}{\partial s}+\frac{\partial u}{\partial z}\cdot\frac{\partial z}{\partial s}$$
$$=4x^3y\cdot re^t+(x^4+2yz^3)\cdot 2rse^{-t}$$
$$+3y^2z^2\cdot r^2\sin t.$$

When $r=2$, $s=1$, and $t=0$, we have $x=2$, $y=2$, and $z=0$, so
$$\frac{\partial u}{\partial s}=4\times 2^3\times 2\times 2\times e^0$$
$$+(2^4+2\times 2\times 0^3)\times 2\times 2\times 1\times e^{-0}$$
$$+3\times 2^2\times 0^2\times 2^2\sin 0$$
$$=128+64=192.$$

2.6 The Derivative Rule of Implicit Function

Let the function $y=f(x)$ be determined by
$$F(x,y)=0. \quad (2.5)$$
Next we use Chain Rule introduced in the previous section to derive the derivative formula of $y=f(x)$. Let's take the derivative of both sides of the equation (2.5) with respect to x by the Chain Rule. We obtain
$$\frac{\partial F}{\partial x}\cdot\frac{\mathrm{d}x}{\mathrm{d}x}+\frac{\partial F}{\partial y}\cdot\frac{\mathrm{d}y}{\mathrm{d}x}=0.$$
Solving $\frac{\mathrm{d}y}{\mathrm{d}x}$, we get the formula
$$\frac{\mathrm{d}y}{\mathrm{d}x}=-\frac{\partial F}{\partial x}\bigg/\frac{\partial F}{\partial y}.$$

Example 1 Find $\frac{\mathrm{d}y}{\mathrm{d}x}$, if $x^3+x^2y-10y^4=0$.

Solution Let $F(x,y)=x^3+x^2y-10y^4$, by equation (2.6) we have
$$\frac{\mathrm{d}y}{\mathrm{d}x}=-\frac{\partial F}{\partial x}\bigg/\frac{\partial F}{\partial y}=-\frac{3x^2+2xy}{x^2-40y^3}.$$

Similarly, the partial derivatives of the implicit function of two variables can also be derived by the Chain Rule. $z=f(x,y)$ is an implicit function determined by the equation
$$F(x,y,z)=0. \quad (2.7)$$
Using Chain Rule to take the derivative of equation (2.7), we have

$$\frac{\partial F}{\partial x}\cdot\frac{dx}{dx}+\frac{\partial F}{\partial y}\cdot\frac{\partial y}{\partial x}+\frac{\partial F}{\partial z}\cdot\frac{\partial z}{\partial x}=0.$$

We note that $\frac{\partial y}{\partial x}=0$ (for x, y is a constant), we get

$$\frac{\partial z}{\partial x}=-\frac{\partial F}{\partial x}\Big/\frac{\partial F}{\partial z}. \qquad (2.8)$$

Taking the derivative of the both sides of equation (2.7) with respect to y, we also can obtain:

$$\frac{\partial z}{\partial y}=-\frac{\partial F}{\partial y}\Big/\frac{\partial F}{\partial z}. \qquad (2.9)$$

Example 2 Let the function
$$x^2+y^2+xyz-z^2=0,$$
find $\frac{\partial z}{\partial x}$ and $\frac{\partial z}{\partial y}$.

Solution Let
$$F(x,y,z)=x^2+y^2+xyz-z^2,$$
then
$$\frac{\partial F}{\partial x}=2x+yz,$$
$$\frac{\partial F}{\partial y}=2y+xz,$$
$$\frac{\partial F}{\partial z}=xy-2z.$$

So, by equation (2.8) and (2.9), we have
$$\frac{\partial z}{\partial x}=-\frac{\partial F}{\partial x}\Big/\frac{\partial F}{\partial z}=\frac{2x+yz}{2z-xy},$$
$$\frac{\partial z}{\partial y}=-\frac{\partial F}{\partial y}\Big/\frac{\partial F}{\partial z}=\frac{2y+xz}{2z-xy}.$$

2.7 Local Extremum, Maximum and Minimum

We know that one of the main uses of ordinary derivatives is to find maximum and minimum values. In this section we will take function of two variables as an example and introduce how to use partial derivatives to obtain maximum and minimum of functions of several variables, then find the maximum and minimum values.

2.7 Local Extremum, Maximum and Minimum / 2.7 局部极值，最值

1. Local Extremum

First, look at the hills and valleys in the graph of $f(x,y)$ shown in Figure 2.8. There are two points where their values of function are larger than the other nearly values of nearby functions. The function values at these two points are called the local maximum value of $f(x,y)$, and the larger of these two values is the maximum. Likewise, there are two points where their values of function are smaller than the other nearly values of nearby functions, these two values are called the local minimum value of $f(x,y)$ and the smaller one is the minimum value of $f(x,y)$.

1. 局部极值

先来看图 2.8 所示的函数 $f(x,y)$ 的图形中的"山峰"和"山谷". 可见，存在两个点，其函数值大于附近的函数值，这两点处的函数值称为函数 $f(x,y)$ 的局部极大值，且其中较大者就是 $f(x,y)$ 的最大值. 同样，存在两个点，其函数值小于附近的函数值，这两点处的函数值称为函数 $f(x,y)$ 的局部极小值，其中较小的就是 $f(x,y)$ 的最小值.

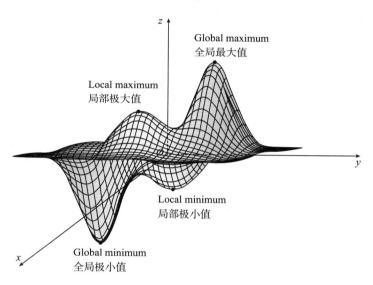

Figure 2.8
图 2.8

Definition 2.5 Let the function of two variables $f(x,y)$ be defined in a neighborhood $U(P_0)$ at point $P_0(x_0,y_0)$. If any point (x,y) within the delated neighborhood $\mathring{U}(P_0)$ can satisfy
$$f(x,y) < f(x_0, y_0)$$
or
$$f(x,y) > f(x_0, y_0).$$
We consider that $f(x,y)$ obtain the **local maximum** or the **local minimum** at point (x_0, y_0), and point (x_0, y_0) is called the **local maximum point** or the **local minimum point**.

The local maximum and the local minimum are referred to as the **local extremum**.

定义 2.5 设二元函数 $f(x,y)$ 在点 $P_0(x_0,y_0)$ 的某个邻域 $U(P_0)$ 内有定义. 如果对于去心邻域 $\mathring{U}(P_0)$ 内任一点 (x,y)，都有
$$f(x,y) < f(x_0, y_0)$$
或
$$f(x,y) > f(x_0, y_0),$$
则称函数 $f(x,y)$ 在点 (x_0, y_0) 处取得**局部极大值**或**局部极小值**，并称点 (x_0, y_0) 为**局部极大值点**或**局部极小值点**.

局部极大值和局部极小值统称为**局部极值**.

Theorem 2.8 (The Necessary Condition of Extremum)

If the function $f(x,y)$ has a local maximum or a local minimum at (x_0, y_0) and its first-order partial derivatives exists there, then
$$f_x(x_0, y_0) = 0 \quad \text{and} \quad f_y(x_0, y_0) = 0.$$

Similar to the function of one variable, a point (x_0, y_0) is called a **stationary point** of $f(x,y)$ if $f_x(x_0, y_0) = 0$ and $f_y(x_0, y_0) = 0$.

The Theorem 2.8 says that if $f(x,y)$ has partial derivative and can obtain a local extremum at (x_0, y_0), and then (x_0, y_0) is a stationary point of $f(x,y)$. However, the stationary point of the function is not necessarily the local extremum of the function. For example, the unique stationary point of $f(x,y) = y^2 - x^2$ is $(0,0)$ but not local extremum point, because $f(x,y)$ can be either positive or negative in any neighborhood at $(0,0)$. In addition, the point where the partial derivative does not exist may also be a local extremum point of the function.

The following theorem gives us a way to determine whether a stationary point is a local extremum point.

Theorem 2.9 (The Sufficient Conditions of Extremum)

Let $f(x,y)$ has the second-order continuous partial derivative in a neighborhood at point (x_0, y_0), and
$$f_x(x_0, y_0) = 0, \quad f_y(x_0, y_0) = 0.$$
Let
$$D = D(x_0, y_0)$$
$$= f_{xx}(x_0, y_0) f_{yy}(x_0, y_0) - [f_{xy}(x_0, y_0)]^2.$$

(1) If $D > 0$ and $f_{xx}(x_0, y_0) > 0$, then $f(x_0, y_0)$ is a local minimum;

(2) If $D > 0$ and $f_{xx}(x_0, y_0) < 0$, then $f(x_0, y_0)$ is a local maximum;

(3) If $D < 0$, then $f(x_0, y_0)$ is not a local extremum.

Remark If $D = 0$, this discriminant is invalid, at this time, we need to determine from the definition whether the stationary point (x_0, y_0) is a local extremum by the definition.

Example 1 Let
$$f(x,y) = x^2 + y^2 - 2x + 6y + 1.$$
Find the local extremum of $f(x,y)$.

Solution We have

定理 2.8（极值的必要条件）

如果函数 $f(x,y)$ 在点 (x_0, y_0) 处有局部极大值或局部极小值，且在该点处的一阶偏导数存在，那么有
$$f_x(x_0, y_0) = 0 \quad 和 \quad f_y(x_0, y_0) = 0.$$

与一元函数类似，如果 $f_x(x_0, y_0) = 0$ 且 $f_y(x_0, y_0) = 0$，那么点 (x_0, y_0) 叫作函数 $f(x,y)$ 的**驻点**.

定理 2.8 说明，如果函数 $f(x,y)$ 在点 (x_0, y_0) 处的偏导数存在且取得局部极值，那么 (x_0, y_0) 是 $f(x,y)$ 的驻点. 但是，函数的驻点不一定是函数的局部极值点. 例如，函数 $f(x,y) = y^2 - x^2$ 的唯一驻点是 $(0,0)$，但是 $(0,0)$ 不是局部极值点，因为在点 $(0,0)$ 的任何一个邻域内，$f(x,y)$ 既可取得正值又可取得负值. 另外，偏导数不存在的点也可能是函数的局部极值点.

下面的定理给我们提供了一种判断驻点是否为局部极值点的方法.

定理 2.9（极值的充分条件）

设函数 $f(x,y)$ 在点 (x_0, y_0) 的某个邻域内具有二阶连续偏导数，并且
$$f_x(x_0, y_0) = 0, \quad f_y(x_0, y_0) = 0.$$
令
$$D = D(x_0, y_0)$$
$$= f_{xx}(x_0, y_0) f_{yy}(x_0, y_0) - [f_{xy}(x_0, y_0)]^2.$$

(1) 如果 $D > 0$，且 $f_{xx}(x_0, y_0) > 0$，那么 $f(x_0, y_0)$ 是局部极小值；

(2) 如果 $D > 0$，且 $f_{xx}(x_0, y_0) < 0$，那么 $f(x_0, y_0)$ 是局部极大值；

(3) 如果 $D < 0$，那么 $f(x_0, y_0)$ 不是局部极值.

注 如果 $D = 0$，该判别法失效，这时我们需要从定义来判定驻点 (x_0, y_0) 是否为局部极值点.

例 1 设函数
$$f(x,y) = x^2 + y^2 - 2x + 6y + 1,$$
求 $f(x,y)$ 的局部极值.

解 我们有

$$f_x(x,y)=2x-2,$$
$$f_y(x,y)=2y+6.$$

Let
$$\begin{cases} f_x(x,y)=0, \\ f_y(x,y)=0, \end{cases}$$
that is
$$\begin{cases} 2x-2=0, \\ 2y+6=0. \end{cases}$$
We could obtain the unique stationary point
$$x=1, \quad y=-3.$$
Thus
$$f_{xx}(x,y)=2, \quad f_{xy}(x,y)=0,$$
$$f_{yy}(x,y)=2,$$
$$D(x,y)=f_{xx}(x,y)f_{yy}(x,y)-[f_{xy}(x,y)]^2$$
$$=2\times 2=4,$$
so we have
$$D(1,-3)=4>0, \quad \text{and} \quad f_{xx}(1,-3)=2>0$$
at the critical point $(1,-3)$. By Theorem 2.9(1), $f(x,y)$ has the local minimum at point $(1,-3)$, and it is
$$f(1,-3)=1^2+(-3)^2-2\times 1+6\times(-3)+1$$
$$=1+9-2-18+1=-9.$$

Example 2 Find the local extremum of
$$f(x,y)=x^4+y^4-4xy+1.$$

Solution We firstly find the stationary points. Because
$$f_x=4x^3-4y, \quad f_y=4y^3-4x,$$
let these two partial derivatives equal to zero, we obtain the equations
$$\begin{cases} x^3-y=0, \\ y^3-x=0. \end{cases}$$
To solve these equations we substitute $y=x^3$ detained from the first equation into the second equation. Then
$$0=x^9-x=x(x^8-1)$$
$$=x(x^4-1)(x^4+1)$$
$$=x(x^2-1)(x^2+1)(x^4+1).$$
So there are three real roots: $x=0,1,-1$. And when $x=0,1,-1, y=0,1,-1$, so the three stationary points are $(0,0),(1,1)$ and $(-1,-1)$.

Next we calculate the second-order partial derivatives and $D(x,y)$:
$$f_{xx}(x,y)=12x^2, \quad f_{xy}(x,y)=-4,$$
$$f_{yy}(x,y)=12y^2,$$

$$D(x,y) = f_{xx}(x,y)f_{yy}(x,y) - [f_{xy}(x,y)]^2$$
$$= 144x^2y^2 - 16.$$

Since
$$D(0,0) = -16 < 0,$$
it follows from the Theorem 2.9(3) that $f(x,y)$ has no extremum at $(0,0)$.

Since
$$D(1,1) = 128 > 0$$
and
$$f_{xx}(1,1) = 12 > 0,$$
we know from the Theorem 2.9(1) that $f(1,1) = -1$ is a local minimum.

Similarly, because
$$D(-1,-1) = 128 > 0$$
and
$$f_{xx}(-1,-1) = 12 > 0,$$
so $f(-1,-1) = -1$ is also a local minimum.

2. Maximum and Minimum

Now we briefly discuss the most value of the function of several variables. Here we still take a function of several variables as an example. We know that when $f(x,y)$ is continuous in a bounded closed region D, $f(x,y)$ must have the maximum or minimum in D. If the maximum (minimum) is taken in the interior of D, the maximum (minimum) is the local maximum(minimum) of $f(x,y)$. So the most point must be extreme point or boundary point. For functions with partial derivatives, the extreme points are the stationary point. Therefore, in practical problems, if it can be determined by the nature of problem that $f(x,y)$ has a the maximum or minimum in D, and $f(x,y)$ has a unique stationary point in D, this stationary point is the most value point of $f(x,y)$. The function value at this point is the maximum or minimum.

Example 3 Find the shortest distance from the point $(1,0,-2)$ to the plane
$$x + 2y + z = 4.$$

Solution The distance from any point (x,y,z) to the point $(1,0,-2)$ is
$$d = \sqrt{(x-1)^2 + y^2 + (z+2)^2}.$$
But if (x,y,z) lies on the plane $x + 2y + z = 4$, then
$$z = 4 - x - 2y.$$

2.7 Local Extremum, Maximum and Minimum

So we have
$$d=\sqrt{(x-1)^2+y^2+(6-x-2y)^2}.$$
Through simplification, we can get
$$d^2 \triangleq f=(x-1)^2+y^2+(6-x-2y)^2.$$
By solving the equations
$$\begin{aligned}f_x &= 2(x-1)-2(6-x-2y)\\&=4x+4y-14=0,\\ f_y &= 2y-4(6-x-2y)\\&=4x+10y-24=0,\end{aligned}$$
we find that the only stationary point is $\left(\dfrac{11}{6},\dfrac{5}{3}\right)$. Because there must be a point on the given plane which is the closest to $(1,0,-2)$, that is, the shortest distance from $(1,0,-2)$ to plane $x+2y+z=4$ exists, or in other words $d^2=f(x,y)$ has minimum. So the stationary point $\left(\dfrac{11}{6},\dfrac{5}{3}\right)$ is the minimum point.

When $x=\dfrac{11}{6}$ and $y=\dfrac{5}{3}$, we have
$$\begin{aligned}d &= \sqrt{(x-1)^2+y^2+(6-x-2y)^2}\\&=\sqrt{\left(\dfrac{5}{6}\right)^2+\left(\dfrac{5}{3}\right)^2+\left(\dfrac{5}{6}\right)^2}\\&=\dfrac{5}{6}\sqrt{6}.\end{aligned}$$
The shortest distance from $(1,0,-2)$ to the plane $x+2y+z=4$ is $\dfrac{5}{6}\sqrt{6}$.

Example 4 Suppose that a rectangular box without a lid is made from $12\ \text{m}^2$ of cardboard. Find the maximum volume of such a box.

Solution Let the length, width and height of the box (in meters) be $x,y,$ and z. Then the volume of the box is
$$V=xyz\quad(x,y,z>0).$$
We can express V as a function of just two variables x and y by using the fact that the area of the four sides and the bottom of the box is
$$2xz+2yz+xy=12.$$
Solving this equation, we can get

所以有
$$d=\sqrt{(x-1)^2+y^2+(6-x-2y)^2}.$$
通过化简,我们得到
$$d^2 \triangleq f=(x-1)^2+y^2+(6-x-2y)^2.$$
通过求解方程组
$$\begin{aligned}f_x &= 2(x-1)-2(6-x-2y)\\&=4x+4y-14=0,\\ f_y &= 2y-4(6-x-2y)\\&=4x+10y-24=0,\end{aligned}$$
我们得到唯一的驻点是 $\left(\dfrac{11}{6},\dfrac{5}{3}\right)$. 因为在给定的平面上必然有一个点最接近$(1,0,-2)$,即点$(1,0,-2)$到平面 $x+2y+z=4$ 的最短距离存在,即函数 $d^2=f(x,y)$ 存在最小值,所以驻点 $\left(\dfrac{11}{6},\dfrac{5}{3}\right)$ 就是最小值点.

当 $x=\dfrac{11}{6},y=\dfrac{5}{3}$ 时,有
$$\begin{aligned}d &= \sqrt{(x-1)^2+y^2+(6-x-2y)^2}\\&=\sqrt{\left(\dfrac{5}{6}\right)^2+\left(\dfrac{5}{3}\right)^2+\left(\dfrac{5}{6}\right)^2}\\&=\dfrac{5}{6}\sqrt{6},\end{aligned}$$
所以点$(1,0,-2)$到平面 $x+2y+z=4$ 的最短距离是 $\dfrac{5}{6}\sqrt{6}$.

例 4 设一个没有盖子的长方形盒子是用 $12\ \text{m}^2$ 的纸板做成的,计算这个盒子的最大容积.

解 设盒子的长、宽、高分别是 x,y,z(单位:m),则盒子的容积是
$$V=xyz\quad(x,y,z>0).$$
我们可以把 V 表示为两个变量 x 和 y 的函数,因为盒子的四个侧面和底的面积满足方程
$$2xz+2yz+xy=12.$$
解这个方程,得到

$$z = \frac{12-xy}{2(x+y)},$$

so the expression for V becomes

$$V = xy\frac{12-xy}{2(x+y)}$$
$$= \frac{12xy - x^2y^2}{2(x+y)}.$$

We compute the partial derivatives of V:

$$\frac{\partial V}{\partial x} = \frac{y^2(12-2xy-x^2)}{2(x+y)^2},$$
$$\frac{\partial V}{\partial y} = \frac{x^2(12-2xy-y^2)}{2(x+y)^2}.$$

Let

$$\begin{cases} \frac{\partial V}{\partial x} = 0, \\ \frac{\partial V}{\partial y} = 0, \end{cases}$$

because $x, y > 0$, we have

$$\begin{cases} 12-2xy-x^2 = 0, \\ 12-2xy-y^2 = 0, \end{cases}$$

then

$$x = \pm 2, \quad y \pm 2.$$

Remove $x = -2, y = -2$, we get the unique stationary point $(2,2)$.

According to the practical significance, the maximum of V must exist, and there exists a unique stationary point $(2,2)$, so the point $(2,2)$ is the maximum point of V. And when $x = 2, y = 2$, there is

$$z = \frac{12-2\times 2}{2(2+2)} = 1,$$

so the maximal capacity of the box is

$$V_{\max} = xyz\big|_{(2,2,1)}$$
$$= 2 \times 2 \times 1$$
$$= 4 \text{ (unit: m}^3\text{)}.$$

Remark We can calculate V_{xx}, V_{xy}, V_{yy}, and use Theorem 2.9 to prove that the point $(2,2)$ is the local maximum point of V, and then we can determine the maximum point of V as $(2,2)$.

Exercises 2
习题 2

1. Let $f(x,y) = \dfrac{y}{x} + xy$. Find the following function values:

   (1) $f(1,2)$; (2) $f\left(\dfrac{1}{4}, 4\right)$;

   (3) $f\left(4, \dfrac{1}{4}\right)$; (4) $f(a,a)$;

   (5) $f\left(\dfrac{1}{x}, x^2\right)$; (6) $f(1,0)$.

2. Let $g(x,y,z) = x^2 \sin yz + xy$, find the following function values:

   (1) $g(1, \pi, 2)$; (2) $g\left(2, 1, \dfrac{\pi}{6}\right)$;

   (3) $g\left(4, 2, \dfrac{\pi}{4}\right)$; (4) $g(\pi, \pi, \pi)$.

3. Sketch the graphes of the following functions:

   (1) $f(x,y) = 6$;
   (2) $f(x,y) = 6 - x$;
   (3) $f(x,y) = 6 - x - 2y$;
   (4) $f(x,y) = 6 - x^2$;
   (5) $f(x,y) = \sqrt{16 - x^2 - y^2}$;
   (6) $f(x,y) = \sqrt{16 - 4x^2 - y^2}$;
   (7) $f(x,y) = 3 - x^2 - y^2$.

4. Find $F(f(t), g(t))$, where

   (1) $F(x,y) = xy^2$, and
   $$f(t) = t^2 \sin t, \quad g(t) = \cos^2 t;$$
   (2) $F(x,y) = xe^2 + y^2$, and
   $$f(t) = \ln t^2, \quad g(t) = e^t.$$

5. Find the domains of the following functions and sketch their graphes:

   (1) $f(x,y) = \sqrt{x+y}$;
   (2) $f(x,y) = \ln(9 - x^2 - 9y^2)$;
   (3) $f(x,y) = \sqrt{1-x^2} - \sqrt{1-y^2}$;

(4) $f(x,y)=\dfrac{\sqrt{y-x^2}}{1-x^2}$;

(5) $f(x,y,z)=\sqrt{1-x^2-y^2-z^2}$.

6. Find the domains of the following functions:

(1) $f(x,y)=(y-2x)^2$;

(2) $f(x,y)=y-\ln x$;

(3) $f(x,y)=y\mathrm{e}^x$;

(4) $f(x,y)=\sqrt{y^2-x^2}$.

7. Suppose that $\lim\limits_{(x,y)\to(3,1)}f(x,y)=6$. Can you find the value of $f(3,1)$? Discussing the continuity of $f(x,y)$.

8. Let
$$f(x,y)=\dfrac{x^2y^3+x^3y^2-5}{2-xy}.$$
Using the expression of $f(x,y)$ to guess the limit of $f(x,y)$ as $(x,y)\to(0,0)$. Then explain why your guess is correct.

9. Find the following limits:

(1) $\lim\limits_{(x,y)\to(5,-2)}(x^5+4x^3y-5xy^2)$;

(2) $\lim\limits_{(x,y)\to(2,1)}\dfrac{4-xy}{x^2+3y^2}$;

(3) $\lim\limits_{(x,y)\to(0,0)}\dfrac{y^4}{x^4+3y^4}$;

(4) $\lim\limits_{(x,y)\to(0,0)}\dfrac{xy\cos y}{3x^2+y^2}$;

(5) $\lim\limits_{(x,y)\to(0,0)}\dfrac{xy}{\sqrt{x^2+y^2}}$;

(6) $\lim\limits_{(x,y)\to(0,0)}\dfrac{x^2y\mathrm{e}^y}{x^4+4y^2}$;

(7) $\lim\limits_{(x,y)\to(0,0)}\dfrac{x^2+y^2}{\sqrt{x^2+y^2+1}-1}$;

(8) $\lim\limits_{(x,y,z)\to(3,0,1)}\mathrm{e}^{-xy}\sin\dfrac{\pi z}{2}$;

(9) $\lim\limits_{(x,y,z)\to(0,0,0)}\dfrac{xy+yz^2+xz^2}{x^2+y^2+z^4}$;

(10) $\lim\limits_{\substack{x\to 1\\ y\to 2}}\dfrac{x\sqrt{x+y^2}}{x+y}$;

(11) $\lim\limits_{\substack{x\to 0\\ y\to 0}}(x+y)\cos\dfrac{1}{xy}$;

(12) $\lim\limits_{\substack{x\to\infty\\ y\to\infty}}\dfrac{1}{x^2+y^2}$;

(13) $\lim\limits_{\substack{x\to 0\\y\to 0}}\dfrac{xy}{\sqrt{xy+1}-1}$;

(14) $\lim\limits_{\substack{x\to 0\\y\to 0}}\dfrac{\sin(xy)}{x(y+1)}$;

(15) $\lim\limits_{\substack{x\to 0\\y\to 0}}\dfrac{y\sin x}{3-\sqrt{x\sin y+9}}$.

10. Draw a graph of function
$$f(x,y)=\dfrac{2x^2+3xy+4y^2}{3x^2+5y^2}$$
by using a computer, and explain why the following limit does not exist:
$$\lim_{(x,y)\to(0,0)}\dfrac{2x^2+3xy+4y^2}{3x^2+5y^2}.$$

11. Show that the following limits are not exist:

(1) $\lim\limits_{\substack{x\to 0\\y\to 0}}\dfrac{x+y}{x-2y}$;

(2) $\lim\limits_{\substack{x\to 0\\y\to 0}}\dfrac{x^4 y^4}{(x^2+y^4)^3}$.

12. Let $g(t)=t^2+\sqrt{t}$, $f(x,y)=2x+3y-6$. Find $h(x,y)=g(f(x,y))$ and discuss continuity of h.

13. Determine the set of points at which the following functions are continuous:

(1) $f(x,y)=\dfrac{1}{x^2-y}$;

(2) $f(x,y)=\arctan(x+\sqrt{y})$;

(3) $f(x,y)=\ln(x^2+y^2-4)$;

(4) $f(x,y,z)=\dfrac{\sqrt{y}}{x^2+y^2+z^2}$;

(5) $f(x,y)=\begin{cases}\dfrac{x^2 y^2}{2x^2+y^2}, & (x,y)\neq(0,0),\\ 0, & (x,y)=(0,0);\end{cases}$

(6) $f(x,y)=\begin{cases}\dfrac{\sin xy}{y}, & y\neq 0,\\ 0, & y=0.\end{cases}$

14. Find all first partial derivatives of the following functions:

(1) $f(x,y)=(2x-y)^4$;

(2) $f(x,y)=(4x-y^2)^{3/2}$;

(3) $f(x,y)=\dfrac{x^2-y^2}{xy}$;

Exercises 2
习题 2

(13) $\lim\limits_{\substack{x\to 0\\y\to 0}}\dfrac{xy}{\sqrt{xy+1}-1}$;

(14) $\lim\limits_{\substack{x\to 0\\y\to 0}}\dfrac{\sin(xy)}{x(y+1)}$;

(15) $\lim\limits_{\substack{x\to 0\\y\to 0}}\dfrac{y\sin x}{3-\sqrt{x\sin y+9}}$.

10. 利用计算机画出函数
$$f(x,y)=\dfrac{2x^2+3xy+4y^2}{3x^2+5y^2}$$
的图形,并解释为什么下面的极限不存在:
$$\lim_{(x,y)\to(0,0)}\dfrac{2x^2+3xy+4y^2}{3x^2+5y^2}.$$

11. 说明下列极限不存在:

(1) $\lim\limits_{\substack{x\to 0\\y\to 0}}\dfrac{x+y}{x-2y}$;

(2) $\lim\limits_{\substack{x\to 0\\y\to 0}}\dfrac{x^4 y^4}{(x^2+y^4)^3}$.

12. 设 $g(t)=t^2+\sqrt{t}$, $f(x,y)=2x+3y-6$,求 $h(x,y)=g(f(x,y))$,并讨论 h 的连续性.

13. 确定下列函数在哪些点集上是连续的:

(1) $f(x,y)=\dfrac{1}{x^2-y}$;

(2) $f(x,y)=\arctan(x+\sqrt{y})$;

(3) $f(x,y)=\ln(x^2+y^2-4)$;

(4) $f(x,y,z)=\dfrac{\sqrt{y}}{x^2+y^2+z^2}$;

(5) $f(x,y)=\begin{cases}\dfrac{x^2 y^2}{2x^2+y^2}, & (x,y)\neq(0,0),\\ 0, & (x,y)=(0,0);\end{cases}$

(6) $f(x,y)=\begin{cases}\dfrac{\sin xy}{y}, & y\neq 0,\\ 0, & y=0.\end{cases}$

14. 求出下列函数的所有一阶偏导数:

(1) $f(x,y)=(2x-y)^4$;

(2) $f(x,y)=(4x-y^2)^{3/2}$;

(3) $f(x,y)=\dfrac{x^2-y^2}{xy}$;

(4) $f(x,y) = e^x \cos y$;
(5) $f(x,y) = e^y \sin x$;
(6) $f(x,y) = (3x^2 + y^2)^{-1/3}$;
(7) $f(x,y) = \sqrt{x^2 - y^2}$;
(8) $f(x,y) = e^{xy}$;
(9) $f(x,y) = \ln(x^2 - y^2)$;
(10) $f(x,y) = y\cos(x^2 + y^2)$;
(11) $f(x,y) = 2\sin x \cos y$.

15. Verify that the following functions satisfies
$$\frac{\partial^2 f}{\partial x \partial y} = \frac{\partial^2 f}{\partial y \partial x}.$$
(1) $f(x,y) = 2x^2 y^3 - x^3 y^5$;
(2) $f(x,y) = (x^3 + y^2)^5$;
(3) $f(x,y) = 3e^{2x} \cos y$;
(4) $f(x,y) = \arctan xy$.

16. If the function $f(x) = \dfrac{2x-y}{xy}$, find $f_x(3,-2)$ and $f_y(3,-2)$.

17. If the function $f(x,y) = \ln(x^2 + xy + y^2)$, find $f_x(-1,4)$ and $f_y(-1,4)$.

18. If the function $f(x,y,z) = (x^3 + y^2 + z)^4$, find
(1) $f_x(x,y,z)$;
(2) $f_y(0,1,1)$;
(3) $f_{xy}(x,y,z)$.

19. Find $\dfrac{dw}{dt}$ by using the Chain Rule, and express your final answer in terms of t.
(1) $w = x^2 y^3, x = t^3, y = t^2$;
(2) $w = x^2 y - y^2 x, x = \cos t, y = \sin t$;
(3) $w = e^x \sin y + e^y \sin x, x = 3t, y = 2t$;
(4) $w = \ln \dfrac{x}{y}, x = \tan t, y = \sec^2 t$;
(5) $w = \sin(xyz^2)$,
$x = t^3, y = t^2, z = t$;
(6) $w = xy + yz + xz$,
$x = t^2, y = 1 - t^2, z = 1 - t$.

20. Find $\dfrac{dw}{dt}$ by using the Chain Rule, and express your final answer in terms of s and t.

(1) $w=x^2y, x=st, y=s-t$;

(2) $w=x^2-y\ln x, x=\dfrac{s}{t}, y=s^2t$;

(3) $w=e^{x^2+y^2}, x=s\sin t, y=t\sin s$;

(4) $w=\ln(x+y)-\ln(x-y)$, $x=te^s, y=e^{st}$;

(5) $w=\sqrt{x^2+y^2+z^2}$, $x=\cos st, y=\sin st, z=s^2$;

(6) $w=e^{xy+z}, x=s+t, y=s-t, z=t^2$.

21. If $z=x^2y, x=2t+s$, and $y=1-st^2$, find $\left.\dfrac{\partial z}{\partial t}\right|_{\substack{s=1\\t=-2}}$.

22. If $z=xy+x+y$, where $x=r+s+t$, and $y=rst$, find $\left.\dfrac{\partial z}{\partial s}\right|_{\substack{r=1\\s=1\\t=2}}$.

23. Suppose that $y(x)$ is determined by the following equations, find $\dfrac{dy}{dx}$:

(1) $x^3+2x^2y-y^3=0$;

(2) $ye^{-x}+5x-7=0$;

(3) $x\sin y+y\cos x=0$;

(4) $x^2\cos y-y^2\sin x=0$.

24. If $3x^2z+y^3-xyz^3=0$, find $\dfrac{\partial z}{\partial x}$.

25. Find the stagnation points of the following functions and indicate whether each such point gives a local maximum or a local minimum, or does not take the extremum.

(1) $f(x,y)=x^2+4y^2-4x$;

(2) $f(x,y)=x^2+4y^2-2x+8y-1$;

(3) $f(x,y)=2x^4-x^2+3y^2$;

(4) $f(x,y)=xy^2-6x^2-3y^2$;

(5) $f(x,y)=xy$;

(6) $f(x,y)=x^3+4y^3-4xy$;

(7) $f(x,y)=xy+\dfrac{2}{x}+\dfrac{4}{y}$;

(8) $f(x,y)=e^{-(x^2+y^2-4y)}$;

(9) $f(x,y)=\cos x+\cos y+\cos(x+y)$
$\left(0<x<\dfrac{\pi}{2}, 0<y<\dfrac{\pi}{2}\right)$;

(10) $f(x,y)=x^2+a^2-2ax\cos y$
$(-\pi<y<\pi)$.

26. Find the local extremum value of following $f(x,y)$ on S:
(1) $f(x,y)=3x+4y$,
   $S=\{(x,y)\mid 0\leqslant x\leqslant 1,-1\leqslant y\leqslant 1\}$;
(2) $f(x,y)=x^2+y^2$,
   $S=\{(x,y)\mid -3\leqslant x\leqslant -1,-1\leqslant y\leqslant 4\}$;
(3) $f(x,y)=x^2-y^2+1$,
   $S=\{(x,y)\mid x^2+y^2\leqslant 1\}$;
(4) $f(x,y)=x^2-6x+y^2-8y+7$,
   $S=\{(x,y)\mid x^2+y^2\leqslant 1\}$.

27. Express a positive number N as a sum of three positive numbers such that the product of these three numbers is a minimum.

28. By using the methods of this section, find the shortest distance from origin to the plane
$$x+2y+3z=12.$$

29. A rectangular box, whose edges are parallel to the coordinate axes, is inscribed in the ellipsoid
$$96x^2+4y^2+4z^2=36.$$
So what is the maximum size of such a box?

30. Find the total differentials of the following functions:
(1) $z=\dfrac{x}{y}$;
(2) $z=\sin(x^2+y^2)$;
(3) $z=\dfrac{y}{\sqrt{x^2+y^2}}$;
(4) $z=\dfrac{x+y}{x-y}$;
(5) $z=e^{y/x}$;
(6) $z=\arcsin\dfrac{x}{y}$ $(y>0)$;
(7) $u=\ln\sqrt{x^2+y^2+z^2}$;
(8) $u=x^{yz}$.

31. Find the full increment and total differential of $z=\dfrac{y}{x}$, when $x=2, y=1, \Delta x=0.1, \Delta y=-0.2$.

32. Find the total differential of $z=e^{xy}$, when $x=1, y=1, \Delta x=0.15, \Delta y=0.1$.

Chapter 3　Double Integral
第 3 章　二重积分

In this chapter, we will introduce the double integral on closed rectangles at first. And then, we will introduce the iterated integral and the double integral on closed non-rectangular regions. Last, the double integral in polar coordinate and can obtain applications of the double integral are provided. We use the double integral to find the volume of general solid, the area of general surface, and the center of mass of laminas and solids of variable density.

在这一章中,我们首先介绍在闭矩形区域上的二重积分;然后,介绍累次积分和在非闭矩形区域上的二重积分;最后,介绍极坐标下的二重积分以及二重积分的应用:用二重积分求立体的体积、物体的表面积、变密度薄板的质心、变密度实体的质心.

3.1　The Double Integral on Closed Rectangles
3.1　闭矩形区域上的二重积分

The main content of calculus are differential and integral. We have studied differential in two-dimensional and three-dimensional space. Let's start to learn the integral in two-dimensional space: the double integral. The theory and the applications of single integral are generalized to multiple integrals.

微积分的主要内容是微分和积分. 我们已经学习了二维平面和三维空间上的微分,下面开始学习二维平面上的积分——二重积分,将一元积分的理论及应用推广到多元积分上.

The intimate connection between integral and differential is enunciated in the fundamental theorems of calculus. This theorem provide the principal theoretical tools for evaluating definite integral. Here we reduce double integral to a succession of single integral, and use definite integral to calculate double integral.

微分和积分的关系由微积分基本定理联系起来,这一理论为计算定积分提供了理论工具. 我们可以把二重积分化成累次积分的形式,并利用定积分来计算二重积分.

Another point worth recalling is the definition of the integral on the interval(definite integral): we form a partition p of the interval $[a,b]$ into subintervals of length $\Delta x_i (i=1,2,\cdots,n)$, pick a sample point $\overline{x}_i$ from the subinterval, note $\|p\|=\max\limits_{1\leqslant i\leqslant n}\{\Delta x_i\}$ and then

另一个值得回顾的是一元函数在区间上的积分——定积分的定义:构造一个分割 p,将区间 $[a,b]$ 分割成长度为 $\Delta x_i (i=1,2,\cdots,n)$ 的 n 个小区间,在第 i 个小区间内任取一点 $\overline{x}_i$,记 $\|p\|=\max\limits_{1\leqslant i\leqslant n}\{\Delta x_i\}$,则

$$\int_a^b f(x)\mathrm{d}x = \lim_{\|p\|\to 0}\sum_{i=1}^n f(\overline{x}_i)\Delta x_i.$$

$$\int_a^b f(x)\mathrm{d}x = \lim_{\|p\|\to 0}\sum_{i=1}^n f(\overline{x}_i)\Delta x_i.$$

We could easily apply this method to the function of two variables $f(x,y)$.

1. The Definition of the Double Integral

Let R be a rectangle with sides parallel to the coordinate axes, that is
$$R=\{(x,y)\,|\,a\leqslant x\leqslant b, c\leqslant y\leqslant d\}.$$
The function $f(x,y)$ is continuous on R, and $f(x,y)\geqslant 0$.

Divide R into n small rectangles $R_k (k=1,2,\cdots,n)$ with the straight lines parallel to the x-axis or the y-axis, as shown in Figure 3.1. We write this split as p, and write that Δx_k and Δy_k are the length and width of each small rectangle. Then $\Delta A_k = \Delta x_k \Delta y_k$ is the area of the small rectangle R_k. In each small rectangle R_k, we choose any point $(\overline{x}_k, \overline{y}_k)$, summation to get the following formula.

$$\sum_{k=1}^{n} f(\overline{x}_k, \overline{y}_k) \Delta A_k. \qquad (3.1)$$

1. 二重积分的定义

设 R 为一个各边均平行于坐标轴的闭矩形区域,即
$$R=\{(x,y)\,|\,a\leqslant x\leqslant b, c\leqslant y\leqslant d\},$$
函数 $f(x,y)$ 在 R 上连续,且 $f(x,y)\geqslant 0$.

如图 3.1 所示,用平行于 x 轴和 y 轴的直线将 R 分割为 n 个小矩形区域 $R_k(k=1, 2,\cdots,n)$,把这一分割记为 p. 分别记 Δx_k 和 Δy_k 为各个小矩形的长和宽,则 $\Delta A_k = \Delta x_k \Delta y_k$ 为小矩形区域 R_k 的面积. 在每个小矩形区域 R_k 内,任选一个样点 $(\overline{x}_k, \overline{y}_k)$,作和式

$$\sum_{k=1}^{n} f(\overline{x}_k, \overline{y}_k) \Delta A_k. \qquad (3.1)$$

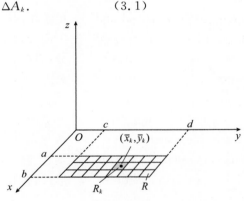

Figure 3.1
图 3.1

This sum formula represents the volume of n small cuboids. It is approximation of the volume V of a curved cylinder[①] which the bottom is R, the top is
$$z=f(x,y).$$

这个和式就代表 n 个小长方体的体积,它是以 R 为底,曲面
$$z=f(x,y)$$
为顶的曲顶柱体[①]的体积 V 的近似值,且 R 被

---

① Curved cylinder refers to such a solid: the solid that the bottom is closed area on the plane, the side is cylinder, the top is surface.

① 曲顶柱体是指这样一种立体:以平面上的有界闭区域为底,侧面为柱面,顶为曲面的立体.

The more detailed R is divided, the more the value of the equation obtained is lifted from the true volume, as shown in Figure 3.2 (V_k, which bottom is small rectangular area R_k, height is $f(\overline{x}_k, \overline{y}_k)$, is the volume of small cylinder in Figure 3.3).

分割得越细,得到的和式的值与真实体积 V 越接近,如图 3.2 (V_k 为以小矩形区域 R_k 为底,$f(\overline{x}_k, \overline{y}_k)$ 为高的小柱体的体积,如图 3.3 所示).

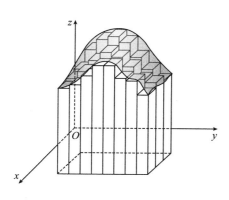

Figure 3.2
图 3.2

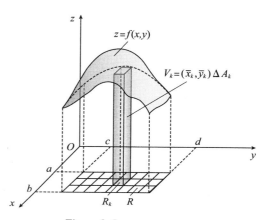

Figure 3.3
图 3.3

We are ready for a formal definition of the double integral. Before giving this definition, we explain the norm $\|p\|$ of split p, which represents the length of the longest diagonal of these n small rectangular areas.

Definition 3.1 Let $f(x, y)$ be a function of two variable that is defined on a closed rectangle R. Doing the above division on R, we get the sum formula (3.1). If the limit of (3.1)

$$\lim_{\|p\| \to 0} \sum_{k=1}^{n} f(\overline{x}_k, \overline{y}_k) \Delta A_k$$

exists, we say that $f(x, y)$ is **integrable** on R, and the limit is the **double integral** of $f(x, y)$ on R, denoted as $\iint\limits_{R} f(x, y) \mathrm{d}A$. That is

$$\iint\limits_{R} f(x, y) \mathrm{d}A = \lim_{\|p\| \to 0} \sum_{k=1}^{n} f(\overline{x}_k, \overline{y}_k) \Delta A_k,$$

where $f(x, y)$ is **integrand**, $f(x, y) \mathrm{d}A$ is the **integral expression**, $\mathrm{d}A$ is the **area element**, x, y are the **integral variables**, R is the **integral area**,

$$\sum_{k=1}^{n} f(\overline{x}_k, \overline{y}_k) \Delta A_k$$

我们已经为给出一个关于二重积分的正式定义做好准备. 在给出这一定义之前, 先解释一下分割 p 的范数 $\|p\|$, 它表示分成的 n 个小矩形区域中对角线最长的那个值.

定义 3.1 设 $f(x, y)$ 为定义在一个闭矩形区域 R 上的二元函数. 对 R 做上述分割, 得到和式 (3.1). 如果和式 (3.1) 的极限

$$\lim_{\|p\| \to 0} \sum_{k=1}^{n} f(\overline{x}_k, \overline{y}_k) \Delta A_k$$

存在, 那么就称 $f(x, y)$ 在 R 上是**可积**的, 并称该极限值为 $f(x, y)$ 在 R 上的**二重积分**, 记为 $\iint\limits_{R} f(x, y) \mathrm{d}A$, 即

$$\iint\limits_{R} f(x, y) \mathrm{d}A = \lim_{\|p\| \to 0} \sum_{k=1}^{n} f(\overline{x}_k, \overline{y}_k) \Delta A_k,$$

其中 $f(x, y)$ 称为**被积函数**, $f(x, y) \mathrm{d}A$ 称为**积分表达式**, $\mathrm{d}A$ 称为**面积元素**, x, y 称为**积分变量**, R 称为**积分区域**,

$$\sum_{k=1}^{n} f(\overline{x}_k, \overline{y}_k) \Delta A_k$$

is the **integral sum.**

This definition of the double integral contains the limits as $\|p\| \to 0$. This is not a limit in the sense of Chapter 1, so we should clarify what this really means. We say that

$$\lim_{\|p\| \to 0} \sum_{k=1}^{n} f(\overline{x}_k, \overline{y}_k) \Delta A_k = L,$$

if for every $\varepsilon > 0$ there exists a $\delta > 0$ such that when each of the split p of the rectangle R made by straight lines parallel to the x- and y-axes satisfies $\|p\| < \delta$, for any choice of the sample points $(\overline{x}_k, \overline{y}_k)$ in the kth rectangle, we have

$$\left| \sum_{k=1}^{n} f(\overline{x}_k, \overline{y}_k) \Delta A_k - L \right| < \varepsilon.$$

Recall that if $f(x) \geqslant 0$, $\int_a^b f(x) \mathrm{d}x$ represents the area of the region surrounded by $y = f(x)$, straight lines $x = a, x = b$, and x-axis. The same in the double integral, if $f(x, y) \geqslant 0$, $\iint_R f(x, y) \mathrm{d}A$ represents the volume of carved cylinder whose bottom is the rectangle R, top is the surface $z = f(x, y)$ (Figure 3.4).

称为积分和.

在二重积分的定义中包含了当 $\|p\| \to 0$ 时的极限概念. 这个极限不同于与第 1 章的极限概念, 我们应该明确其真正的含义. 我们说

$$\lim_{\|p\| \to 0} \sum_{k=1}^{n} f(\overline{x}_k, \overline{y}_k) \Delta A_k = L,$$

是指如果对于任给 $\varepsilon > 0$, 存在 $\delta > 0$, 使得当矩形区域 R 的每一个由平行于 x 轴和 y 轴的直线所做的分割 p 都满足 $\|p\| < \delta$ 时, 在第 k 个小矩形区域 R_k 内任取一点 $(\overline{x}_k, \overline{y}_k)$, 都有

$$\left| \sum_{k=1}^{n} f(\overline{x}_k, \overline{y}_k) \Delta A_k - L \right| < \varepsilon$$

成立.

在定积分中, 如果 $f(x) \geqslant 0$, 那么 $\int_a^b f(x) \mathrm{d}x$ 代表的是由曲线 $y = f(x)$ 与直线 $x = a, x = b$ 及 x 轴所围成曲边梯形的面积. 同样, 在二重积分中, 如果 $f(x, y) \geqslant 0$, 那么 $\iint_R f(x, y) \mathrm{d}A$ 代表的是以矩形区域 R 为底、曲面 $z = f(x, y)$ 为顶的曲顶柱体的体积 (图 3.4).

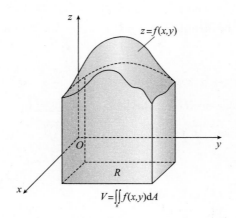

Figure 3.4
图 3.4

If $f(x,y) \leqslant 0$, above curved cylinder is below the Oxy-plane, then $\iint\limits_R f(x,y)\mathrm{d}A$ express the negative value of the volume of curved cylinder. If $f(x,y)$ is positive in some areas of R, while in the rest of the area is negative, then $\iint\limits_R f(x,y)\mathrm{d}A$ is equal to the algebraic sum of the volume of curved cylinder corresponding to each area. The volume of the cylinder above the Oxy-plane takes a positive value, and the volume of the cylinder below the Oxy-plane takes a negative value.

2. The Existence Question of Double Integral

As the function of one variable, not any function of two variables is integrable on a given closed rectangle R. In particular, a function that is unbounded on R will always fail to be integrable.

Theorem 3.1 (Integrability Theorem) If $f(x,y)$ is bounded on the closed rectangle R and continuous there except on a finite number of smooth curves, then $f(x,y)$ is integrable on R.

In particular, if $f(x,y)$ is continuous on the whole closed rectangle R, then $f(x,y)$ is integrable there.

As a consequence, most of the common functions are integrable on every closed rectangle. For example,
$$f(x,y) = \mathrm{e}^{\sin(xy)} - y^3 \cos(x^2 y)$$
is integrable on every closed rectangle. On the other hand,
$$g(x,y) = \frac{x^2 y - 2x}{y - x^2}$$
would fail to be integrable on any closed rectangle which intersected the parabola $y = x^2$. The staircase function of Figure 3.5 is integrable on closed rectangle R because it is discontinuities occur along line segment. Unless otherwise specified, it is assumed that the double integral of the $f(x,y)$ we will discuss exists.

如果 $f(x,y) \leqslant 0$，上述曲顶柱体在 Oxy 平面下方，那么 $\iint\limits_R f(x,y)\mathrm{d}A$ 表示曲顶柱体体积的负值；如果 $f(x,y)$ 在 R 的部分区域上是正的，而在其余的部分区域上是负的，那么 $\iint\limits_R f(x,y)\mathrm{d}A$ 等于各部分区域对应曲顶柱体体积的代数和，其中在 Oxy 平面上方的柱体体积取正值，在 Oxy 平面下方的柱体体积取负值.

2. 二重积分的存在性问题

如同一元函数那样，不是每个二元函数都在给定的闭矩形区域 R 上可积. 特别地，在 R 上无界的函数是不可积的.

定理 3.1（可积性定理） 如果函数 $f(x,y)$ 在闭矩形区域 R 上有界，并且除了有限条光滑曲线外，$f(x,y)$ 是连续的，那么 $f(x,y)$ 在 R 上可积.

特别地，如果 $f(x,y)$ 在整个闭矩形区域 R 上是连续的，则 $f(x,y)$ 在其上是可积的.

因此，大多数函数在闭矩形区域上都可积. 例如，函数
$$f(x,y) = \mathrm{e}^{\sin(xy)} - y^3 \cos(x^2 y)$$
在任意的闭矩形区域上都可积. 然而，函数
$$g(x,y) = \frac{x^2 y - 2x}{y - x^2}$$
在任何与抛物线 $y = x^2$ 相交的闭矩形区域上是不可积. 如图 3.5 中所示的阶梯形函数，它在闭矩形区域 R 上是可积的，因为这个函数只在线段上是不连续的. 以下除特殊说明外，假定我们所讨论的函数 $f(x,y)$ 的二重积分都存在.

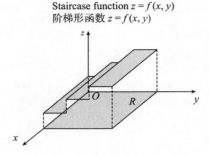

Figure 3.5

图 3.5

3. Properties of the Double Integral

The double integral inherits most of the properties of the single integral.

Property 1 The double integral has linear properties, that is

(1) $\iint\limits_R k f(x,y) \mathrm{d}A = k \iint\limits_R f(x,y) \mathrm{d}A$;

(2) $\iint\limits_R [f(x,y) + g(x,y)] \mathrm{d}A$
$= \iint\limits_R f(x,y) \mathrm{d}A + \iint\limits_R g(x,y) \mathrm{d}A.$

Property 2 The double integral is additive on closed rectangle where only the boundaries overlap. That is to say if closed rectangle $R = R_1 \cup R_2$, and R_1, R_2 only overlap on the border, then (Figure 3.6)

$\iint\limits_R f(x,y) \mathrm{d}A$
$= \iint\limits_{R_1} f(x,y) \mathrm{d}A + \iint\limits_{R_2} f(x,y) \mathrm{d}A.$

3. 二重积分的性质

二重积分继承了定积分大部分的性质.

性质 1 二重积分具有线性性质, 即

(1) $\iint\limits_R k f(x,y) \mathrm{d}A = k \iint\limits_R f(x,y) \mathrm{d}A$;

(2) $\iint\limits_R [f(x,y) + g(x,y)] \mathrm{d}A$
$= \iint\limits_R f(x,y) \mathrm{d}A + \iint\limits_R g(x,y) \mathrm{d}A.$

性质 2 二重积分在仅有边界重叠的闭矩形区域上具有可加性(图 3.6), 即若闭矩形区域 $R = R_1 \cup R_2$, 且闭矩形区域 R_1 与 R_2 只在边界重叠, 则

$\iint\limits_R f(x,y) \mathrm{d}A$
$= \iint\limits_{R_1} f(x,y) \mathrm{d}A + \iint\limits_{R_2} f(x,y) \mathrm{d}A.$

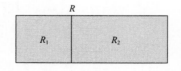

Figure 3.6

图 3.6

For the double integral, the comparative nature still exists.

对于二重积分, 比较性质依然存在.

Property 3 If $f(x,y) \leqslant g(x,y)$ for all $(x,y) \in R$, then
$$\iint\limits_R f(x,y)\mathrm{d}A \leqslant \iint\limits_R g(x,y)\mathrm{d}A.$$

In particular, because of $-|f(x,y)| \leqslant f(x,y) \leqslant |f(x,y)|$, we have
$$\left|\iint\limits_R f(x,y)\mathrm{d}A\right| \leqslant \iint\limits_R |f(x,y)|\,\mathrm{d}A.$$

We note that if $f(x,y)=1$ on the closed rectangle R then the double integral of $f(x,y)$ on R is the area of R. Therefore,
$$\iint\limits_R k\,\mathrm{d}A = k\iint\limits_R 1\,\mathrm{d}A = kA(R),$$
where $A(R)$ is the area of R.

Property 4 (Valuation Inequality) Let M, m be the maximum and minimum of the continuous function $f(x,y)$ on closed rectangle R, then we have
$$mA(R) \leqslant \iint\limits_R f(x,y)\mathrm{d}A \leqslant MA(R).$$

Property 5 (Integral Median Theorem) If function is continuous on R, there is at least one point (ζ, η) on R, such that
$$\iint\limits_R f(x,y)\mathrm{d}A = f(\zeta,\eta)A(R).$$

4. Simple Calculation of Double Integrals

This topic will receive major attention in the next section, where we will develop a powerful tool for calculating double integrals. However, we can already calculate some simple double integrals, and we can approximate others.

Example 1 Let
$$f(x,y) = \begin{cases} 1, & 0 \leqslant x \leqslant 3, 0 \leqslant y \leqslant 1, \\ 2, & 0 \leqslant x \leqslant 3, 1 < y \leqslant 2, \\ 3, & 0 \leqslant x \leqslant 3, 2 < y \leqslant 3. \end{cases}$$

The $f(x,y)$ is the staircase function of Figure 3.5. Calculate $\iint\limits_R f(x,y)\mathrm{d}A$, where
$$R = \{(x,y) \mid 0 \leqslant x \leqslant 3, 0 \leqslant y \leqslant 3\}.$$

Solution Introduce closed rectangles R_1, R_2 and R_3, as follows:
$$R_1 = \{(x,y) \mid 0 \leqslant x \leqslant 3, 0 \leqslant y \leqslant 1\},$$
$$R_2 = \{(x,y) \mid 0 \leqslant x \leqslant 3, 1 \leqslant y \leqslant 2\},$$
$$R_3 = \{(x,y) \mid 0 \leqslant x \leqslant 3, 2 \leqslant y \leqslant 3\}.$$

Then, using the additivity property of the double integral, we obtain
$$\iint_R f(x,y) \mathrm{d}A$$
$$= \iint_{R_1} f(x,y) \mathrm{d}A + \iint_{R_2} f(x,y) \mathrm{d}A$$
$$+ \iint_{R_3} f(x,y) \mathrm{d}A$$
$$= A(R_1) + 2A(R_2) + 3A(R_3)$$
$$= 3 + 2 \times 3 + 3 \times 3 = 18.$$

In this derivation, we also use the fact that the value of $f(x,y)$ on the boundary of the closed rectangle does not affect the value of the integral.

Example 1 is just a simple example, and to be honest we cannot do much more without more tools. However, we can always approximate a double integral by calculating integral sum. In general, we can expect the approximation to be better by the thinner the partition we use.

Example 2 Find the approximate value of the double integral $\iint_R f(x,y) \mathrm{d}A$, here
$$f(x,y) = \frac{64 - 8x + y^2}{16},$$
and
$$R = \{(x,y) \mid 0 \leqslant x \leqslant 4, 0 \leqslant y \leqslant 8\}.$$
When calculating, we divide R into eight equal square areas and take the center of each square as a sample point $(\overline{x}_k, \overline{y}_k)$, then use the integral sum to find.

Solution The values of the function at the required sample points are as follows in Figure 3.7:

(1) $(\overline{x}_1, \overline{y}_1) = (1,1)$,
$$f(\overline{x}_1, \overline{y}_1) = f(1,1) = \frac{57}{16};$$

解 引入如下闭矩形区域 R_1, R_2 和 R_3:
$$R_1 = \{(x,y) \mid 0 \leqslant x \leqslant 3, 0 \leqslant y \leqslant 1\},$$
$$R_2 = \{(x,y) \mid 0 \leqslant x \leqslant 3, 1 \leqslant y \leqslant 2\},$$
$$R_3 = \{(x,y) \mid 0 \leqslant x \leqslant 3, 2 \leqslant y \leqslant 3\}.$$

然后,利用二重积分的可加性,得到
$$\iint_R f(x,y) \mathrm{d}A$$
$$= \iint_{R_1} f(x,y) \mathrm{d}A + \iint_{R_2} f(x,y) \mathrm{d}A$$
$$+ \iint_{R_3} f(x,y) \mathrm{d}A$$
$$= A(R_1) + 2A(R_2) + 3A(R_3)$$
$$= 3 + 2 \times 3 + 3 \times 3 = 18.$$

在这个推导过程中,我们利用了 $f(x,y)$ 在闭矩形区域边界上的取值不影响积分值这个事实.

例 1 只是一个很简单的例子,但实际上,在还没有更多工具的情况下,我们什么也做不了. 然而,用求积分和的方法可以求二重积分的近似值. 更一般地,我们也可以采用更细的分割来取得更精确的近似值.

例 2 求二重积分 $\iint_R f(x,y) \mathrm{d}A$ 的近似值,这里
$$f(x,y) = \frac{64 - 8x + y^2}{16},$$
且
$$R = \{(x,y) \mid 0 \leqslant x \leqslant 4, 0 \leqslant y \leqslant 8\}.$$
计算时把 R 分成 8 个相等的正方形区域,取各个正方形的中心为样点 $(\overline{x}_k, \overline{y}_k)$,然后用积分和来求.

解 如图 3.7 所示,要求的样点及相关的函数值如下:

(1) $(\overline{x}_1, \overline{y}_1) = (1,1)$,
$$f(\overline{x}_1, \overline{y}_1) = f(1,1) = \frac{57}{16};$$

(2) $(\bar{x}_2, \bar{y}_2) = (1, 3)$,
$$f(\bar{x}_2, \bar{y}_2) = f(1, 3) = \frac{65}{16};$$
(3) $(\bar{x}_3, \bar{y}_3) = (1, 5)$,
$$f(\bar{x}_3, \bar{y}_3) = f(1, 5) = \frac{81}{16};$$
(4) $(\bar{x}_4, \bar{y}_4) = (1, 7)$,
$$f(\bar{x}_4, \bar{y}_4) = f(1, 7) = \frac{105}{16};$$
(5) $(\bar{x}_5, \bar{y}_5) = (3, 1)$,
$$f(\bar{x}_5, \bar{y}_5) = f(3, 1) = \frac{41}{16};$$
(6) $(\bar{x}_6, \bar{y}_6) = (3, 3)$,
$$f(\bar{x}_6, \bar{y}_6) = f(3, 3) = \frac{49}{16};$$
(7) $(\bar{x}_7, \bar{y}_7) = (3, 5)$,
$$f(\bar{x}_7, \bar{y}_7) = f(3, 5) = \frac{65}{16};$$
(8) $(\bar{x}_8, \bar{y}_8) = (3, 7)$,
$$f(\bar{x}_8, \bar{y}_8) = f(3, 7) = \frac{89}{16}.$$

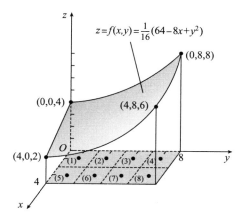

Figure 3.7

图 3.7

Thus, since $\Delta A_k = 4$,
$$\iint_R f(x, y) \, dA$$
$$\approx \sum_{k=1}^{8} f(\bar{x}_k, \bar{y}_k) \Delta A_k$$

$$= 4\sum_{k=1}^{8} f(\overline{x}_k, \overline{y}_k)$$
$$= \frac{4(57+65+81+105+41+49+65+89)}{16}$$
$$= 138.$$

3.2 Iterated Integrals
3.2 累次积分

Now we must face the most important problem, that is calculating $\iint_R f(x,y)\, dA$, where R is the closed rectangle:

$$R = \{(x,y) \mid a \leqslant x \leqslant b, c \leqslant x \leqslant d\}.$$

1. Change the Double Integral to the Iterated Integral

Suppose that $f(x,y) \geqslant 0$ on R so that we may interpret the double integral $\iint_R f(x,y)$ as the volume V of the curved cylinder whose bottom is R, top is $z = f(x,y)$ (Figure 3.8), that is

$$V = \iint_R f(x,y)\, dA. \qquad (3.2)$$

现在我们必须面对最重要的问题了,那就是计算 $\iint_R f(x,y)\, dA$ 的值,这里 R 是闭矩形区域:

$$R = \{(x,y) \mid a \leqslant x \leqslant b, c \leqslant y \leqslant d\}.$$

1. 化二重积分为累次积分

若在 R 上满足 $f(x,y) \geqslant 0$,我们就可以把二重积分 $\iint_R f(x,y)$ 理解为以 R 为底,曲面 $z = f(x,y)$ 为顶的曲顶柱体的体积 V 了(图 3.8),即

$$V = \iint_R f(x,y)\, dA. \qquad (3.2)$$

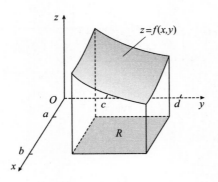

Figure 3.8
图 3.8

There is another way to calculate the volume of the curved cylinder showed by Figure 3.8. This way at least intuitively seems just as valid. Use plane $y = C$(constant) to cut the curved cylinder into a sheet parallel to Ozx-plane. The typical example of this sheet is shown in Figure 3.9(a). The area of the sheet depends on how far it is from the Ozx-plane. That is to say the area of sheet depends on y. Therefore, we denote this area by $A(y)$ (Figure 3.9(b)). At the moment, the volume ΔV of the sheet is approximately

$$\Delta V \approx A(y)\Delta y.$$

From the learned knowledge: the micro element method (subdivision, taking approximate value, integral), we may write

$$V = \int_c^d A(y)\,\mathrm{d}y.$$

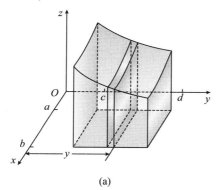
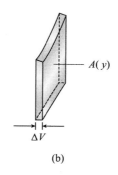

(a) (b)

Figure 3.9
图 3.9

In fact, this is the volume of this solid whose sectional area has known. The sectional area is $A(y)$.

On the other hand, for fixed y, we may calculate the sectional area $A(y)$ by means of an ordinary single integral (Figure 3.10), that is

$$A(y) = \int_a^b f(x,y)\,\mathrm{d}x.$$

Then we conclude that

$$V = \int_c^d A(y)\,\mathrm{d}y$$
$$= \int_c^d \left[\int_a^b f(x,y)\,\mathrm{d}x\right]\mathrm{d}y. \qquad (3.3)$$

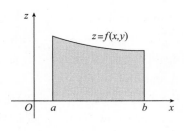

Figure 3.10
图 3.10

The last expression is called an **iterated integral**. It is said we should find the integral on x in $[a,b]$ first while y is treated as a constant, then find the integral on y in $[c,d]$.

Both equations (3.2) and (3.3) represent the volume of curved cylinder. Therefore we could obtain the result we want:

$$\iint\limits_R f(x,y)\,\mathrm{d}A = \int_c^d \left[\int_a^b f(x,y)\,\mathrm{d}x\right]\mathrm{d}y.$$

If we had begun the process above by slicing the curved cylinder with plane $x=C$ (constant) parallel to the Oyz-plane, we would have obtained another iterated integral, with the integrations occurring in the opposite order:

$$\iint\limits_R f(x,y)\,\mathrm{d}A = \int_a^b \left[\int_c^d f(x,y)\,\mathrm{d}y\right]\mathrm{d}x. \quad (3.4)$$

This iterated integral is gotten by finding the integral on y in $[c,d]$ first while x is treated as a constant, then finding the integral on x in $[a,b]$.

There are two remarks. First, while the two formulas are derived under the assumption that $f(x,y)$ is nonnegative, they are valid in general. Second, the whole process of integrating double integral into iterated integral would be rather pointless unless iterated integrals can be evaluated. Fortunately, as long as the appropriate order of integral is selected, iterated integrals are often easy to evaluate, as we demonstrate next.

2. Calculating Iterated Integral

Let's start with a simple example.

Example 1 Find the iterated integral

上式右端这个表达式称为**累次积分**,它表示先把 y 看作常数,对 x 在 $[a,b]$ 上求定积分,再对 y 在 $[c,d]$ 上求定积分.

由于方程(3.2)和(3.3)均表示曲顶柱体的体积 V,因此可以得到我们想要的结果:

$$\iint\limits_R f(x,y)\,\mathrm{d}A = \int_c^d \left[\int_a^b f(x,y)\,\mathrm{d}x\right]\mathrm{d}y.$$

如果我们在开始的时候采用平行于 Oyz 平面的平面 $x=C$(常数)切割这个曲顶柱体,则可以得到另外一个积分顺序相反的累次积分:

$$\iint\limits_R f(x,y)\,\mathrm{d}A = \int_a^b \left[\int_c^d f(x,y)\,\mathrm{d}y\right]\mathrm{d}x. \quad (3.4)$$

这是先把 x 看作常数,对 y 在 $[c,d]$ 上求定积分,再对 x 在 $[a,b]$ 上求定积分的累次积分.

这里有两个值得注意的地方:第一,虽然上面两个公式是在 $f(x,y)$ 非负的前提下得到的,但是它们是普遍适用的. 第二,如果累次积分不能计算,那整个将二重积分化为累次积分的过程将是无意义的. 幸运的是,只要选择适当的积分次序,累次积分通常是容易计算的,像我们下面所展示的那样.

2. 累次积分的计算

我们从一个简单的例子开始.

例 1 计算累次积分

$$\int_0^3 \left[\int_1^2 (2x+3y)\,dx\right] dy.$$

Solution First, we find the integral on x in $[1,2]$ while y is treated as a constant, so

$$\int_1^2 (2x+3y)\,dx = \left[x^2+3yx\right]_1^2$$
$$= 4+6y-(1+3y)$$
$$= 3+3y.$$

Consequently,

$$\int_0^3 \left[\int_1^2 (2x+3y)\,dx\right] dy = \int_0^3 (3+3y)\,dy$$
$$= \left[3y+\frac{3}{2}y^2\right]_0^3$$
$$= 9+\frac{27}{2} = \frac{45}{2}.$$

Example 2 Find the iterated integral

$$\int_1^2 \left[\int_0^3 (2x+3y)\,dy\right] dx.$$

Solution Note that we have simply reversed the order of integration of Example 1. From the geometric meaning of the double integral, we know that the values of these two integrals are the same. Below to verify the result. Because

$$\int_0^3 (2x+3y)\,dy = \left[2xy+\frac{3}{2}y^2\right]_0^3$$
$$= 6x+\frac{27}{2},$$

then

$$\int_1^2 \left[\int_0^3 (2x+3y)\,dy\right] dx = \int_1^2 \left(6x+\frac{27}{2}\right) dx$$
$$= \left[3x^2+\frac{27}{2}x\right]_1^2$$
$$= 12+27-\left(3+\frac{27}{2}\right)$$
$$= \frac{45}{2}.$$

From now on, we usually omit the brackets in the iterated integral. For example, write

$$\int_c^d \left[\int_a^b f(x,y)\,dx\right] dy$$

as

$$\int_c^d \int_a^b f(x,y)\,dx\,dy.$$

Remark The order of dx and dy is important because it specifies which integration is to be done first. The first integration of the iterated integral involves the integrand $f(x,y)$, and the nearest integral symbol to the left of it and the first symbol dx or dy to the right of it. We will sometimes refer to this integral as the **inner integral** and value as the **inner integration**.

Example 3 Calculate the iterated integral
$$\int_0^8 \int_0^4 \frac{1}{16}(64 - 8x + y^2)dxdy.$$

Solution Note that this iterated integral corresponds to the double integral of Example 2 of Section 3.1. We usually consider the inner integral separately and calculate from inside to outside:
$$\int_0^8 \int_0^4 \frac{1}{16}(64 - 8x + y^2)dxdy$$
$$= \frac{1}{16}\int_0^8 \left[64x - 4x^2 + xy^2\right]_0^4 dy$$
$$= \frac{1}{16}\int_0^8 (256 - 64 + 4y^2)dy$$
$$= \int_0^8 \left(12 + \frac{1}{4}y^2\right)dy$$
$$= \left[12y + \frac{y^3}{12}\right]_0^8$$
$$= 96 + \frac{512}{12} = 138\frac{2}{3}.$$

Now we can calculate volumes for a wide variety of solids. Let's consider a specific examples.

Example 4 Find the volume V of the curved cylinder whose top is the surface $z = 4 - x^2 - y$ and bottom is the closed rectangle $R = \{(x,y) \mid 0 \leqslant x \leqslant 1, 0 \leqslant y \leqslant 2\}$ (Figure 3.11).

Solution According to the geometric meaning of the double integral, the required volume is
$$V = \iint\limits_R (4 - x^2 - y)dA$$
$$= \int_0^2 \int_0^1 (4 - x^2 - y)dxdy$$
$$= \int_0^2 \left[4x - \frac{x^3}{3} - yx\right]_0^1 dy$$
$$= \int_0^2 \left(4 - \frac{1}{3} - y\right)dy$$

注 dx 和 dy 的顺序是很重要的,因为它能决定对被积函数的积分先后顺序. 累次积分的第一个积分包括被积函数 $f(x,y)$ 和它左边最近的积分符号以及它右边第一个符号 dx 或 dy. 我们有时称这个积分为**内层积分**,并把它的值称为**内层积分值**.

例 3 计算累次积分
$$\int_0^8 \int_0^4 \frac{1}{16}(64 - 8x + y^2)dxdy.$$

解 注意到这个累次积分与3.1节中例2的二重积分相对应. 我们通常将内层积分分开考虑,从里到外计算:
$$\int_0^8 \int_0^4 \frac{1}{16}(64 - 8x + y^2)dxdy$$
$$= \frac{1}{16}\int_0^8 \left[64x - 4x^2 + xy^2\right]_0^4 dy$$
$$= \frac{1}{16}\int_0^8 (256 - 64 + 4y^2)dy$$
$$= \int_0^8 \left(12 + \frac{1}{4}y^2\right)dy$$
$$= \left[12y + \frac{y^3}{12}\right]_0^8$$
$$= 96 + \frac{512}{12} = 138\frac{2}{3}.$$

现在我们可以计算各种各样立体的体积了. 下面来看一个具体的例子.

例 4 求以闭矩形区域 $R = \{(x,y) \mid 0 \leqslant x \leqslant 1, 0 \leqslant y \leqslant 2\}$ 为底,曲面 $z = 4 - x^2 - y$ 为顶的曲顶柱体的体积(图 3.11).

解 根据二重积分的几何意义,所求的体积为
$$V = \iint\limits_R (4 - x^2 - y)dA$$
$$= \int_0^2 \int_0^1 (4 - x^2 - y)dxdy$$
$$= \int_0^2 \left[4x - \frac{x^3}{3} - yx\right]_0^1 dy$$
$$= \int_0^2 \left(4 - \frac{1}{3} - y\right)dy$$

$$= \left[\frac{11}{3}y - \frac{1}{2}y^2\right]_0^2$$
$$= \frac{22}{3} - 2 = \frac{16}{3}.$$

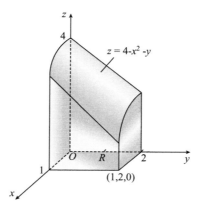

Figure 3.11
图 3.11

3.3 The Double Integral on Non Closed Rectangular Regions
3.3 非闭矩形区域上的二重积分

1. The Definition of Double Integral on a Bounded Closed Area

Consider any bounded closed area S which is surrounded by a closed rectangle R (Figure 3.12). Suppose that $f(x,y)$ is defined on R, and $f(x,y)=0$ on the part of R outside of S (Figure 3.13). Then $f(x,y)$ is **integrable** on S if it is integrable on R and written as

$$\iint\limits_S f(x,y)\mathrm{d}A = \iint\limits_R f(x,y)\mathrm{d}A.$$

We call it the double integral of $f(x,y)$ on S.

It is known from the property of double integral of $f(x,y)$ on the closed rectangle R, we can easily find that the double integral on the bounded closed area S also has the same properties: (1) Linear nature; (2) Additivity of bounded closed areas with smooth curve over lapping; (3) Comparative nature; (4) Valuation Inequality; (5) Integral

1. 有界闭区域上的二重积分的定义

考虑平面上的任意有界闭区域 S,它被闭矩形区域 R 所包围(图 3.12).假设 $f(x,y)$ 在 R 上有定义,且在 R 上 S 之外有 $f(x,y)=0$ (图 3.13).如果 $f(x,y)$ 在 R 上可积,则我们说 $f(x,y)$ 在 S 上**可积**,记为

$$\iint\limits_S f(x,y)\mathrm{d}A = \iint\limits_R f(x,y)\mathrm{d}A,$$

并称之为 $f(x,y)$ 在 S 上的二重积分.

由 $f(x,y)$ 在闭矩形区域 R 上的二重积分的性质可知,我们容易得到 $f(x,y)$ 在有界闭区域 S 上的二重积分也具有同样的性质:(1) 线性性质;(2) 在仅沿光滑曲线交叠的有界闭区域上的可加性;(3) 比较性质;(4) 估值不等式;(5) 积分中值定理(参见 3.1 节).

Figure 3.12

图 3.12

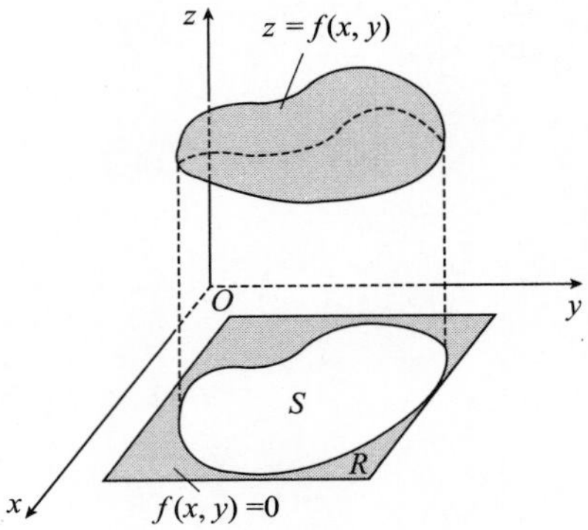

Figure 3.13

图 3.13

Median Theorem (see Section 3.1).

2. Calculation of Double Integral on a Bounded Closed Area

Sets with curved boundaries can be very complicated. For our purposes, it will be sufficient to consider x-simple area and y-simple area because the general bounded closed region can be divided into a finite area of the two regions. An area S is x-simple if it is simple in the y-direction, meaning that a line in this direction intersects S in a single interval(or point or not at all). Thus, an area S is x-**simple** (Figure 3.14) if there are functions φ_1 and φ_2 on $[a,b]$ such that
$$S=\{(x,y)\mid \varphi_1(x)\leqslant y\leqslant \varphi_2(x), a\leqslant x\leqslant b\}.$$
An area S is y-**simple** (Figure 3.15) if there are functions ψ_1 and ψ_2 on $[c,d]$ such that
$$S=\{(x,y)\mid \psi_1(y)\leqslant x\leqslant \psi_2(y), c\leqslant y\leqslant d\}.$$

For example, Figure 3.16 exhibits an area that is neither x-simple nor y-simple.

Now suppose that we will calculate the double integral $\iint\limits_{S} f(x,y)\,dA$ of a function $f(x,y)$ on a x-simple area S. We enclose S by a closed rectangle R (Figure 3.17) and

2. 有界闭区域上的二重积分的计算

平面曲线围成的区域相当复杂. 对于我们来说,只考虑 x 型区域和 y 型区域就已经足够了,因为一般的有界闭区域都可以划分为这两种区域的有限个并集. 如果用一条平行于 y 轴的直线穿过区域 S,它与 S 的交集是一线段 (或者一个点或空集),即存在 $[a,b]$ 上满足条件
$$S=\{(x,y)\mid \varphi_1(x)\leqslant y\leqslant \varphi_2(x), a\leqslant x\leqslant b\}$$
的函数 φ_1 和 φ_2,则称区域 S 为 x **型区域**(图 3.14);同样,如果存在 $[c,d]$ 上满足条件
$$S=\{(x,y)\mid \psi_1(y)\leqslant x\leqslant \psi_2(y), c\leqslant y\leqslant d\}$$
的函数 ψ_1 和 ψ_2 (图 3.15),那么称区域 S 为 y **型区域**.

例如,图 3.16 给出的区域既不是 x 型区域,也不是 y 型区域.

现在假设我们要计算函数 $f(x,y)$ 在 x 型区域 S 上的二重积分 $\iint\limits_{S} f(x,y)\,dA$. 我们用一个闭矩形区域 R 围住 S (图 3.17),并令

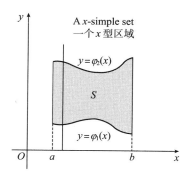

Figure 3.14

图 3.14

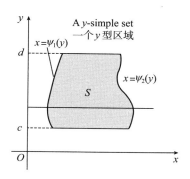

Figure 3.15

图 3.15

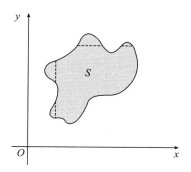

Figure 3.16

图 3.16

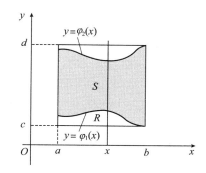

Figure 3.17

图 3.17

let $f(x,y)=0$ in $R\backslash S$. Then
$$\iint_S f(x,y)\,dA = \iint_R f(x,y)\,dA$$
$$= \int_a^b \left[\int_c^d f(x,y)\,dy\right]dx,$$
$$= \int_a^b \left[\int_{\varphi_1(x)}^{\varphi_2(x)} f(x,y)\,dy\right]dx,$$

that is
$$\iint_S f(x,y)\,dA = \int_a^b \int_{\varphi_1(x)}^{\varphi_2(x)} f(x,y)\,dy\,dx.$$

In the inner integration of the right end of above formula, x is fixed. Thus, this integration is along the vertical line of Figure 3.17. So when $f(x,y) \geq 0$, this integration yields the area $A(x)$ of the cross section shown in Figure 3.18. $A(x)$ is integrated from a to b. It is obviously that

在 $R\backslash S$ 上有 $f(x,y)=0$，则有
$$\iint_S f(x,y)\,dA = \iint_R f(x,y)\,dA$$
$$= \int_a^b \left[\int_c^d f(x,y)\,dy\right]dx$$
$$= \int_a^b \left[\int_{\varphi_1(x)}^{\varphi_2(x)} f(x,y)\,dy\right]dx,$$

即
$$\iint_S f(x,y)\,dA = \int_a^b \int_{\varphi_1(x)}^{\varphi_2(x)} f(x,y)\,dy\,dx.$$

在上式右端的内层积分中，x 是保持固定的，因而这一积分沿着图 3.17 中 x 轴的垂线方向进行。于是，当 $f(x,y) \geq 0$ 时，此积分得出图 3.18 所示的截面的面积 $A(x)$。再对 $A(x)$ 从 a 到 b 积分，显然就是以 S 为底，$z=$

the volume of the curved cylinder whose bottom is S and top is $z = f(x, y)$, that is
$$V = \iint_S f(x, y) \, dA.$$

If the area S is y-simple (Figure 3.15), similar reasoning leads to the formula
$$\iint_S f(x, y) \, dA = \int_c^d \int_{\psi_1(y)}^{\psi_2(y)} f(x, y) \, dx \, dy.$$

If the area S is neither x-simple nor y-simple (Figure 3.16), it usually be considered as a union of pieces that have one or the other of these properties. For example, the annular of Figure 3.19 is not simple in either direction, but it is the union of the four x-simple sets S_1, S_2, S_3, S_4. The double integrals on these pieces can be calculated and added together to obtain the integral on S.

$f(x, y)$为顶的曲顶柱体的体积 V，即
$$V = \iint_S f(x, y) \, dA.$$

若区域 S 为 y 型区域（图 3.15），类似的推理可导出公式
$$\iint_S f(x, y) \, dA = \int_c^d \int_{\psi_1(y)}^{\psi_2(y)} f(x, y) \, dx \, dy.$$

若区域 S 既不是 x 型区域也不是 y 型区域（图 3.16），它通常可看作若干个 x 型区域和 y 型区域的组合. 例如，图 3.19 中的环形物既不是 x 型区域也不是 y 型区域，但它是 4 个 x 型区域 S_1, S_2, S_3, S_4 的组合. 这些分块上的二重积分可以计算，加起来就可以得到 S 上的二重积分.

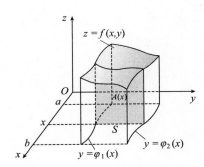

Figure 3.18

图 3.18

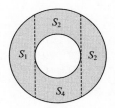

Figure 3.19

图 3.19

Example 1 Calculate the iterated integral
$$\int_3^5 \int_{-x}^{x^2} (4x + 10y) \, dy \, dx.$$

Solution We first perform the inner integration with respect to y, temporarily thinking of x as constant (see Figure 3.20), and obtain
$$\int_3^5 \int_{-x}^{x^2} (4x + 10y) \, dy \, dx$$
$$= \int_3^5 \left[4xy + 5y^2 \right]_{-x}^{x^2} dx$$
$$= \int_3^5 \left[(4x^3 + 5x^4) - (-4x^2 + 5x^2) \right] dx$$

例 1 计算累次积分
$$\int_3^5 \int_{-x}^{x^2} (4x + 10y) \, dy \, dx.$$

解 我们首先完成关于 y 的内层积分，暂时把 x 看作常数（图 3.20），得到
$$\int_3^5 \int_{-x}^{x^2} (4x + 10y) \, dy \, dx$$
$$= \int_3^5 \left[4xy + 5y^2 \right]_{-x}^{x^2} dx$$
$$= \int_3^5 \left[(4x^3 + 5x^4) - (-4x^2 + 5x^2) \right] dx$$

$$= \int_3^5 (5x^4 + 4x^3 - x^2)\,dx$$
$$= \left[x^5 + x^4 - \frac{x^3}{3}\right]_3^5$$
$$= \frac{10\,180}{3} = 3393\frac{1}{3}.$$

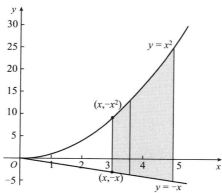

Figure 3.20

图 3.20

Remark For the iterated integral, the outer integral cannot have limits that depend on any integral variable.

Example 2 Calculate the iterated integral
$$\int_0^1 \int_0^{y^2} 2y e^x \,dx\,dy$$

Solution The region of integration is shown in Figure 3.21. It is y-simple, so we have
$$\int_0^1 \int_0^{y^2} 2y e^x \,dx\,dy = \int_0^1 \left[\int_0^{y^2} 2y e^x \,dx\right] dy$$
$$= \int_0^1 \left[2y e^x\right]_0^{y^2} dy$$
$$= \int_0^1 (2y e^{y^2} - 2y e^0)\,dy$$
$$= \int_0^1 e^{y^2} \cdot 2y\,dy - 2\int_0^1 y\,dy$$
$$= \int_0^1 e^{y^2}\,d(y^2) - 2\int_0^1 y\,dy$$
$$= \left[e^{y^2}\right]_0^1 - 2\left[\frac{y^2}{2}\right]_0^1$$
$$= e - 1 - 2 \times \frac{1}{2}$$
$$= e - 2.$$

注 累次积分,其外层积分不能含有依赖于任何积分变量的积分限.

例 2 计算累次积分
$$\int_0^1 \int_0^{y^2} 2y e^x \,dx\,dy.$$

解 积分区域如图 3.21 所示,它是 y 型区域. 于是有
$$\int_0^1 \int_0^{y^2} 2y e^x \,dx\,dy = \int_0^1 \left[\int_0^{y^2} 2y e^x \,dx\right] dy$$
$$= \int_0^1 \left[2y e^x\right]_0^{y^2} dy$$
$$= \int_0^1 (2y e^{y^2} - 2y e^0)\,dy$$
$$= \int_0^1 e^{y^2} \cdot 2y\,dy - 2\int_0^1 y\,dy$$
$$= \int_0^1 e^{y^2}\,d(y^2) - 2\int_0^1 y\,dy$$
$$= \left[e^{y^2}\right]_0^1 - 2\left[\frac{y^2}{2}\right]_0^1$$
$$= e - 1 - 2 \times \frac{1}{2}$$
$$= e - 2.$$

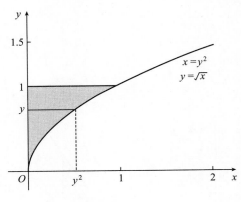

Figure 3.21
图 3.21

We turn to the problem of calculating volumes by means of the iterated integral.

Example 3 Use the double integral to find the volume of the tetrahedron bounded by the coordinate planes and the plane $3x+6y+4z-12=0$.

Solution Using S to represent the projection of tetrahedrons on the Oxy-plane (Figure 3.22 and 3.23).

下面我们转向用累次积分的方法解决体积的计算问题.

例 3 用二重积分计算由坐标面和平面 $3x+6y+4z-12=0$ 所围成的四面体的体积.

解 用 S 表示四面体在 Oxy 平面上的投影(图 3.22 和图 3.23).

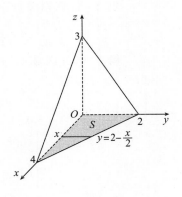

Figure 3.22
图 3.22

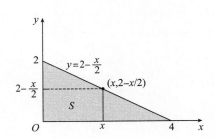

Figure 3.23
图 3.23

The given plane $3x+6y+4z-12=0$ intersects the Oxy-plane in the line
$$x+2y-4=0,$$
a segment of which belongs to the boundary of S. Since this equation can be written as $y=2-\dfrac{x}{2}$ and $x=4-2y$, then

给定平面 $3x+6y+4z-12=0$ 与 Oxy 平面交线是
$$x+2y-4=0,$$
其中的一段属于 S 的边界. 由于直线方程可写作 $y=2-\dfrac{x}{2}$ 和 $x=4-2y$, 所以 S 可看作

S can be thought as the x-simple area
$$S = \left\{(x,y) \,\middle|\, 0 \leqslant x \leqslant 4, 0 \leqslant y \leqslant 2 - \frac{x}{2}\right\}$$
or as the y-simple area
$$S = \{(x,y) \mid 0 \leqslant x \leqslant 4 - 2y, 0 \leqslant y \leqslant 2\}.$$

We will treat S as a x-simple set. The final result would be the same either way, as you should verify. So, the volume V of the solid is
$$V = \iint_S \frac{3}{4}(4 - x - 2y)\,\mathrm{d}A.$$

In order to write the above double integral into the iterated integral, we fix x and integrate along a line from $y=0$ to $y = 2 - \frac{x}{2}$ (Figure 3.22 and 3.23), and then integrate the result from $x=0$ to $x=4$. Thus,
$$\begin{aligned}
V &= \int_0^4 \int_0^{2-\frac{x}{2}} \frac{3}{4}(4 - x - 2y)\,\mathrm{d}y\,\mathrm{d}x \\
&= \int_0^4 \left[\frac{3}{4}\int_0^{2-\frac{x}{2}} (4 - x - 2y)\,\mathrm{d}y\right]\mathrm{d}x \\
&= \int_0^4 \frac{3}{4}\left[4y - xy - y^2\right]_0^{2-\frac{x}{2}}\mathrm{d}x \\
&= \frac{3}{16}\int_0^4 (16 - 8x + x^2)\,\mathrm{d}x \\
&= \frac{3}{16}\left[16x - 4x^2 + \frac{x^3}{3}\right]_0^4 = 4.
\end{aligned}$$

You may recall that the volume of a tetrahedron is one-third the area of the base times the height. In this case,
$$V = \frac{1}{3} \times 4 \times 3 = 4.$$

It confirms that our answer is correct.

Example 4 Find the volume of the solid in the first octant ($x \geqslant 0, y \geqslant 0, z \geqslant 0$) bounded by the circular parabolic $z = x^2 + y^2$, the cylinder $x^2 + y^2 = 4$, and the Oxy-planes (Figure 3.24).

Solution In the Oxy-plane, suppose that the closed area S is bounded by a quarter of the circle $x^2 + y^2 = 4$ and the lines $x = 0$ and $y = 0$.

Obviously, the enclosed solid in the first octant is a curved cylinder whose bottom is S and top is the parabolic

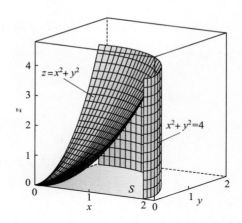

Figure 3.24
图 3.24

$z = x^2 + y^2$. Its volume is
$$V = \iint_S (x^2 + y^2)\, dA.$$

Although S can be viewed as either a x-simple or an y-simple area, we shall treat S as the latter and write its boundary curves as $x = \sqrt{4-y^2}$, $x = 0$ and $y = 0$, that is
$$S = \{(x,y) \mid 0 \leqslant x \leqslant \sqrt{4-y^2}, 0 \leqslant y \leqslant 2\}.$$
Figure 3.25 shows the region S in the Oxy-plane.

体,它的体积为
$$V = \iint_S (x^2 + y^2)\, dA.$$

S 既可看成 x 型区域也可看成 y 型区域,这里假设为后者,且边界曲线为 $x = \sqrt{4-y^2}$, $x = 0$, $y = 0$, 则
$$S = \{(x,y) \mid 0 \leqslant x \leqslant \sqrt{4-y^2}, 0 \leqslant y \leqslant 2\}.$$
图 3.25 给出了 Oxy 平面上的区域 S.

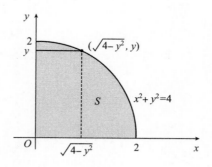

Figure 3.25
图 3.25

Now our goal is to calculate
$$V = \iint_S (x^2 + y^2)\, dA$$
by means of the iterated integral. This time we first fix y and integrate along a line from $x = 0$ to $x = \sqrt{4-y^2}$ (Figure 3.25)

我们的目标是按累次积分的方式计算
$$V = \iint_S (x^2 + y^2)\, dA.$$
这次我们先固定 y,然后沿着直线从 $x = 0$ 到 $x = \sqrt{4-y^2}$ 积分(图 3.25),再把结果从 $y = 0$

and then integrate the result from $y=0$ to $y=2$. Then we have

$$V = \iint_S (x^2 + y^2) dA$$
$$= \int_0^2 \int_0^{\sqrt{4-y^2}} (x^2 + y^2) dx dy$$
$$= \int_0^2 \left[\frac{1}{3}(4-y^2)^{3/2} + y^2\sqrt{4-y^2} \right] dy,$$

By the trigonometric substitution $y = 2\sin\theta$, the latest integral can be rewritten as

$$\int_0^{\frac{\pi}{2}} \left(\frac{8}{3}\cos^3\theta + 8\sin^2\theta\cos\theta \right) 2\cos\theta d\theta$$
$$= \int_0^{\frac{\pi}{2}} \left(\frac{16}{3}\cos^4\theta + 16\sin^2\theta\cos^2\theta \right) d\theta$$
$$= \frac{16}{3} \int_0^{\frac{\pi}{2}} \cos^2\theta (1 - \sin^2\theta + 3\sin^2\theta) d\theta$$
$$= \frac{16}{3} \int_0^{\frac{\pi}{2}} (\cos^2\theta + 2\sin^2\theta\cos^2\theta) d\theta$$
$$= \frac{16}{3} \int_0^{\frac{\pi}{2}} \left(\cos^2\theta + \frac{1}{2}\sin^2 2\theta \right) d\theta$$
$$= \frac{16}{3} \int_0^{\frac{\pi}{2}} \left(\frac{1 + \cos 2\theta}{2} + \frac{1 - \cos 4\theta}{4} \right) d\theta$$
$$= 2\pi,$$

that is, the volume required is $V = 2\pi$.

Is this answer reasonable? Note that the volume of the complete quarter cylinder is $\frac{1}{4}\pi r^2 h = \frac{1}{4}\pi \times 2^2 \times 4 = 4\pi$. One-half of this value is the desired volume.

Example 5 By changing the order of integration, calculate the iterated integral

$$\int_0^4 \int_{x/2}^2 e^{y^2} dy dx.$$

Solution The inner integral cannot be evaluated as it stands, because e^{y^2} does not have an antiderivative in terms of elementary functions. However, we recognize that the given iterated integral is equal to $\iint_S e^{y^2} dA$, where

$$S = \left\{ (x,y) \mid \frac{x}{2} \leq y \leq 2, 0 \leq x \leq 4 \right\}$$
$$= \{(x,y) \mid 0 \leq x \leq 2y, 0 \leq y \leq 2\}$$

(Figure 3.26). If we write this double integral as an iterated integral, we get

$$\int_0^2 \int_0^{2y} e^{y^2} dx dy = \int_0^2 \left[x e^{y^2} \right]_0^{2y} dy$$
$$= \int_0^2 2y e^{y^2} dy$$
$$= \left[e^{y^2} \right]_0^2$$
$$= e^4 - 1.$$

$$\int_0^2 \int_0^{2y} e^{y^2} dx dy = \int_0^2 \left[x e^{y^2} \right]_0^{2y} dy$$
$$= \int_0^2 2y e^{y^2} dy$$
$$= \left[e^{y^2} \right]_0^2$$
$$= e^4 - 1.$$

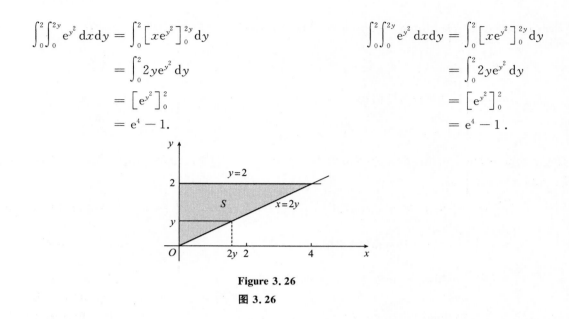

Figure 3.26

图 3.26

3.4 The Double Integral in Polar Coordinates
3.4 极坐标下的二重积分

Such as the circle, the cardioid and the rose, these curves are easier to describe in terms of polar coordinates than in Cartesian coordinates. Thus, we can expect that it is easier to use polar coordinates for the double integral in the area enclosed by such a curve. Now we will study in depth just one particular transformation, from Cartesian coordinates to polar coordinates, because this technique is so useful.

Let area R have the shape shown in Figure 3.27, which is called a **polar rectangle area**. It can be expressed as
$$R = \{(x,y) \mid a^2 \leqslant x^2 + y^2 \leqslant b^2\}$$
or
$$R = \{(r,\theta) \mid a \leqslant r \leqslant b, \alpha \leqslant \theta \leqslant \beta\},$$
where
$$0 \leqslant a \leqslant b \quad \text{and} \quad \beta - \alpha \leqslant 2\pi.$$

像圆、心形、玫瑰形等，这些曲线在极坐标下比在笛卡儿直角坐标下更容易描述. 于是，我们可以预期，在这样的曲线所围成的闭区域上的二重积分使用极坐标更容易计算. 现在我们深入研究一个从直角坐标到极坐标的特殊变换，因为这种技能非常有用.

设区域 R 的图形如图 3.27 所示，我们称 R 是一个**极矩形区域**，它可以表示为
$$R = \{(x,y) \mid a^2 \leqslant x^2 + y^2 \leqslant b^2\}$$
或
$$R = \{(r,\theta) \mid a \leqslant r \leqslant b, \alpha \leqslant \theta \leqslant \beta\},$$
其中
$$0 \leqslant a \leqslant b \quad 且 \quad \beta - \alpha \leqslant 2\pi.$$

3.4 The Double Integral in Polar Coordinates
3.4 极坐标下的二重积分

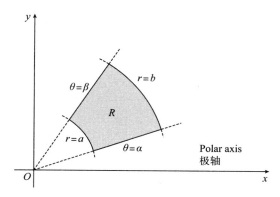

Figure 3.27
图 3.27

Let $z=f(x,y)$ be defined on R, $f(x,y)$ is continuous and nonnegative. Then the volume V of the curved cylinder (Figure 3.28) whose bottom is R and top is surface $z=f(x,y)$ is

$$V = \iint\limits_R f(x,y)\,\mathrm{d}A. \qquad (3.5)$$

设函数 $z=f(x,y)$ 定义在 R 上，$f(x,y)$ 是连续且非负的，则以 R 为底，曲面 $z=f(x,y)$ 为顶的曲顶柱体（图 3.28）的体积 V 为

$$V = \iint\limits_R f(x,y)\,\mathrm{d}A. \qquad (3.5)$$

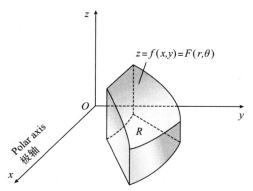

Figure 3.28
图 3.28

Since the transformation formula between Cartesian coordinates and polar coordinates is

$$\begin{cases} x = r\cos\theta, \\ y = r\sin\theta. \end{cases}$$

In polar coordinates, the equation of surface $z=f(x,y)$ can be written as

$$z = f(x,y) = f(r\cos\theta, r\sin\theta) \\ \triangleq F(r,\theta).$$

由于直角坐标与极坐标之间的变换公式为

$$\begin{cases} x = r\cos\theta, \\ y = r\sin\theta. \end{cases}$$

所以在极坐标下，曲面 $z=f(x,y)$ 的方程可以写为

$$z = f(x,y) = f(r\cos\theta, r\sin\theta) \\ \triangleq F(r,\theta).$$

The volume V will be calculated in a new way by using polar coordinates.

Use the polar coordinates network composed by a family of concentric circles ($r=$ constant) and a family of polar rays ($\theta=$ constant) to split R into n small polar rectangular regions $R_1, R_2, \cdots, R_n$. Let Δr_k and $\Delta \theta_k$ denote the side length and the corresponding angle of R_k, as shown in Figure 3.29.

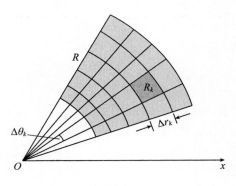

Figure 3.29
图 3.29

By the area formula of sector, we can know the area of R_k is given by
$$A(R_k) = \bar{r}_k \Delta r_k \Delta \theta_k,$$
where $\bar{r}_k$ is the average radius of R_k which is the average of $r_k + \Delta r_k$ and r_k. Thus, we have
$$V \approx \sum_{k=1}^{n} F(\bar{r}_k, \bar{\theta}_k) \bar{r}_k \Delta r_k \Delta \theta_k,$$
where we take the limit as the norm of the partition approaches zero, we can get the actual volume. This limit actually is a double integral, that is,
$$V = \iint_R F(r,\theta) r \, dr \, d\theta$$
$$= \iint_R f(r\cos\theta, r\sin\theta) r \, dr \, d\theta. \quad (3.6)$$

From (3.5) and (3.6), we have
$$\iint_R f(x,y) \, dA = \iint_R f(r\cos\theta, r\sin\theta) r \, dr \, d\theta$$
$$= \iint_R f(r,\theta) r \, dr \, d\theta. \quad (3.7)$$

The integral area is

$$R = \{(r,\theta) \mid a \leqslant r \leqslant b, \alpha \leqslant \theta \leqslant \beta\},$$

so

$$\iint_R f(x,y)\,\mathrm{d}A = \int_\alpha^\beta \int_a^b F(r,\theta) r\,\mathrm{d}r\,\mathrm{d}\theta. \quad (3.8)$$

The above formula is derived under the assumption that $f(x,y)$ is nonnegative, but it is valid for every general function, in particular for continuous functions of arbitrary sign.

Example 1 Find the volume V of the curved cylinder area whose bottom is the polar rectangle
$$R = \{(r,\theta) \mid 0 \leqslant r \leqslant a, 0 \leqslant \theta \leqslant 2\pi\}$$
and top is the surface $z = 2e^{-x^2-y^2}$.

Solution Since $x^2 + y^2 = r^2$, the required volume is
$$V = \iint_R 2e^{-x^2-y^2}\,\mathrm{d}A = 2\iint_R e^{-r^2} r\,\mathrm{d}r\,\mathrm{d}\theta$$
$$= 2\int_0^{2\pi}\left(\int_0^a e^{-r^2} r\,\mathrm{d}r\right)\mathrm{d}\theta = \int_0^{2\pi}\left[-e^{-r^2}\right]_0^a \mathrm{d}\theta$$
$$= (1 - e^{-a^2})\int_0^{2\pi}\mathrm{d}\theta = 2\pi(1 - e^{-a^2}).$$

Similar to the promotion of integral area on Cartesian coordinates from closed rectangle to generally bounded closed area S, the formula (3.7) is established, that is,

$$\iint_S f(x,y)\,\mathrm{d}A = \iint_S f(r\cos\theta, r\sin\theta) r\,\mathrm{d}r\,\mathrm{d}\theta. \quad (3.9)$$

For the area on polar coordinates, we are interested in the following forms: So-called θ-simple set and r-simple set. We say the area S is a θ-**simple set** (Figure 3.30) if S can be expressed as

$$S = \{(r,\theta) \mid \varphi_1(\theta) \leqslant r \leqslant \varphi_2(\theta), \alpha \leqslant \theta \leqslant \beta\}. \quad (3.10)$$

We say the area S is a r-**simple set** (Figure 3.31) if S can be expressed as

$$S = \{(r,\theta) \mid a \leqslant r \leqslant b, \psi_1(r) \leqslant \theta \leqslant \psi_2(r)\}. \quad (3.11)$$

Obviously, any bounded closed area S can be partitioned into a finite number of θ-simple set or r-simple set.

When S is θ-simple set (3.10), the formula (3.9) can be further written as

$$\iint_S f(x,y)\,\mathrm{d}A = \int_\alpha^\beta \int_{\varphi_1(\theta)}^{\varphi_2(\theta)} f(r\cos\theta, r\sin\theta) r\,\mathrm{d}r\,\mathrm{d}\theta.$$

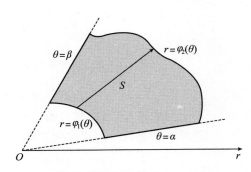

Figure 3.30

图 3.30

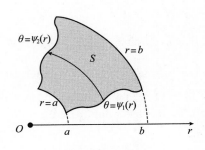

Figure 3.31

图 3.31

When S is r-simple set (3.11), the formula (3.9) can be further written as

$$\iint_S f(x,y)\,\mathrm{d}A = \int_a^b \int_{\psi_1(r)}^{\psi_2(r)} f(r\cos\theta, r\sin\theta) r\,\mathrm{d}\theta\,\mathrm{d}r.$$

Example 2 Calculate the double integral $\iint_S 2y\,\mathrm{d}A$ where S is the region in the first quadrant that is outside the circle $r=2$ and inside the cardioid $r=2(1+\cos\theta)$ (Figure 3.32).

当 S 为 r 型区域(3.11)时，公式(3.9)可进一步写成

$$\iint_S f(x,y)\,\mathrm{d}A = \int_a^b \int_{\psi_1(r)}^{\psi_2(r)} f(r\cos\theta, r\sin\theta) r\,\mathrm{d}\theta\,\mathrm{d}r.$$

例 2 计算二重积分 $\iint_S 2y\,\mathrm{d}A$，这里 S 是在第一象限中圆 $r=2$ 以外心形线 $r=2(1+\cos\theta)$ 以内的区域(图 3.32)。

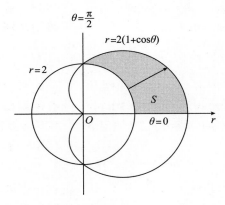

Figure 3.32

图 3.32

Solution Since S is a θ-simple set,

$$S = \left\{(r,\theta) \,\Big|\, 2 \leqslant r \leqslant 2(1+\cos\theta), 0 \leqslant \theta \leqslant \frac{\pi}{2}\right\},$$

we have

解 由于 S 是一个 θ 型区域：

$$S = \left\{(r,\theta) \,\Big|\, 2 \leqslant r \leqslant 2(1+\cos\theta), 0 \leqslant \theta \leqslant \frac{\pi}{2}\right\},$$

我们有

$$\iint_S 2y\,dA = \int_0^{\frac{\pi}{2}} \int_2^{2(1+\cos\theta)} 2(r\sin\theta) r\,dr\,d\theta$$
$$= 2\int_0^{\frac{\pi}{2}} \left[\frac{r^3 \sin\theta}{3}\right]_2^{2(1+\cos\theta)} d\theta$$
$$= \frac{16}{3}\int_0^{\frac{\pi}{2}} [(1+\cos\theta)^3 \sin\theta - \sin\theta]\,d\theta$$
$$= \frac{16}{3}\left[-\frac{1}{4}(1+\cos\theta)^4 + \cos\theta\right]_0^{\frac{\pi}{2}}$$
$$= \frac{16}{3}\left[-\frac{1}{4} + 0 - (-4+1)\right]$$
$$= \frac{44}{3}.$$

Example 3 Find the volume of the solid under the surface $z = x^2 + y^2$, above the Oxy-plane, and inside the cylinder $x^2 + y^2 = 2y$ (Figure 3.33).

例 3 求在曲面 $z = x^2 + y^2$ 下方，Oxy 平面上方以及柱体 $x^2 + y^2 = 2y$ 内的立体体积(图 3.33).

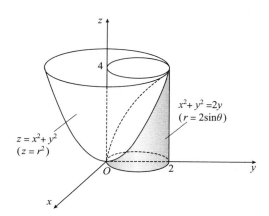

Figure 3.33
图 3.33

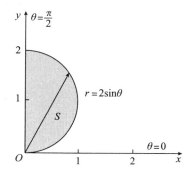

Figure 3.34
图 3.34

Solution By symmetry, we can double the volume in the first octant, and the required volume is given by
$$V = 2\iint_S (x^2+y^2)\,dA,$$
where S is a semi-circular disc as shown in Figure 3.34. Since S could be expressed as
$$S = \left\{(r,\theta) \,\middle|\, 0 \leqslant r \leqslant 2\sin\theta, 0 \leqslant \theta \leqslant \frac{\pi}{2}\right\},$$
and the surface $z = x^2 + y^2$ is $z = r^2$,

解 由对称性，所求体积是在第一卦限内的体积的两倍，所求体积为
$$V = 2\iint_S (x^2+y^2)\,dA,$$
其中 S 是如图 3.34 所示的半圆盘. 由于 S 可表示为
$$S = \left\{(r,\theta) \,\middle|\, 0 \leqslant r \leqslant 2\sin\theta, 0 \leqslant \theta \leqslant \frac{\pi}{2}\right\},$$
而曲面 $z = x^2 + y^2$ 即 $z = r^2$，所以

$$V = 2\int_0^{\frac{\pi}{2}}\int_0^{2\sin\theta} r^2 \cdot r\,dr\,d\theta$$
$$= 2\int_0^{\frac{\pi}{2}} \left[\frac{r^4}{4}\right]_0^{2\sin\theta} d\theta$$
$$= 8\int_0^{\frac{\pi}{2}} \sin^4\theta\,d\theta$$
$$= \frac{3\pi}{2}.$$

Example 4 Prove that
$$I = \int_0^{+\infty} e^{-x^2}\,dx = \frac{\sqrt{\pi}}{2}.$$

Proof We are going to use a circuitous but ingenious way to prove it. First recall that
$$I = \int_0^{+\infty} e^{-x^2}\,dx = \lim_{b\to+\infty}\int_0^b e^{-x^2}\,dx.$$

Now we let V_b be the volume of curved cylinder whose bottom is the square of the vertex $(\pm b, \pm b)$ and top is surface $z = e^{-x^2-y^2}$ (Figure 3.35), then
$$V_b = \int_{-b}^{b}\int_{-b}^{b} e^{-x^2-y^2}\,dy\,dx$$
$$= \int_{-b}^{b} e^{-x^2}\left(\int_{-b}^{b} e^{-y^2}\,dy\right)dx$$
$$= \int_{-b}^{b} e^{-x^2}\,dx \int_{-b}^{b} e^{-y^2}\,dy$$
$$= \left(\int_{-b}^{b} e^{-x^2}\,dx\right)^2$$
$$= 4\left(\int_0^{b} e^{-x^2}\,dx\right)^2.$$

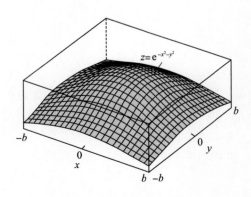

Figure 3.35
图 3.35

The volume of the spatial area below the surface $z = e^{-x^2-y^2}$ and above the Oxy-plane is

$$V = \lim_{b \to +\infty} V_b = \lim_{b \to +\infty} 4\left(\int_0^b e^{-x^2} dx\right)^2$$
$$= 4\left(\int_0^{+\infty} e^{-x^2} dx\right)^2 = 4I^2.$$

On the other hand, we can also calculate V using polar coordinates.

Let S be the disk whose center is origin and radius is b. $\widetilde{V}_b$ is the volume of the curved cylinder whose bottom is S and top is the surface $z = e^{-x^2-y^2}$. Obviously there is

$$\lim_{b \to +\infty} V_b = \lim_{b \to +\infty} \widetilde{V}_b,$$

and

$$S = \{(r,\theta) \mid 0 \leqslant r \leqslant b, 0 \leqslant \theta \leqslant 2\pi\}.$$

So

$$\widetilde{V}_b = \iint_S e^{-x^2-y^2} dA$$
$$= \int_0^{2\pi} \int_0^b e^{-r^2} r dr d\theta.$$

Then

$$V = \lim_{b \to +\infty} V_b = \lim_{b \to +\infty} \widetilde{V}_b$$
$$= \int_0^{2\pi} \left(\int_0^b e^{-r^2} r dr\right) d\theta$$
$$= \lim_{b \to +\infty} \int_0^{2\pi} \left[-\frac{1}{2} e^{-r^2}\right]_0^b d\theta$$
$$= \lim_{b \to +\infty} \frac{1}{2} \int_0^{2\pi} (1 - e^{-b^2}) d\theta$$
$$= \lim_{b \to +\infty} \pi(1 - e^{-b^2}) = \pi.$$

So

$$I = \frac{\sqrt{V}}{2} = \frac{\sqrt{\pi}}{2}.$$

Example 5 Find the double integral

$$I = \iint_S (2 - x^2 - y^2) dA,$$

where $S = \{(x,y) \mid x^2 + y^2 \leqslant 1\}$.

Solution According to the symmetry of S, let S_1 be the part of S in the first quadrant, there is

$$I = \iint_S (2 - x^2 - y^2) dA$$

于是,在曲面 $z = e^{-x^2-y^2}$ 下方和整个 Oxy 平面上方的空间区域的体积为

$$V = \lim_{b \to +\infty} V_b = \lim_{b \to +\infty} 4\left(\int_0^b e^{-x^2} dx\right)^2$$
$$= 4\left(\int_0^{+\infty} e^{-x^2} dx\right)^2 = 4I^2.$$

另一方面,我们利用极坐标来计算 V.

设 S 是以原点为圆心,b 为半径的圆盘,而 $\widetilde{V}_b$ 是以 S 为底,曲面 $z = e^{-x^2-y^2}$ 为顶的曲顶柱体的体积,则显然有

$$\lim_{b \to +\infty} V_b = \lim_{b \to +\infty} \widetilde{V}_b,$$

且

$$S = \{(r,\theta) \mid 0 \leqslant r \leqslant b, 0 \leqslant \theta \leqslant 2\pi\}.$$

所以

$$\widetilde{V}_b = \iint_S e^{-x^2-y^2} dA$$
$$= \int_0^{2\pi} \int_0^b e^{-r^2} r dr d\theta.$$

于是

$$V = \lim_{b \to +\infty} V_b = \lim_{b \to +\infty} \widetilde{V}_b$$
$$= \int_0^{2\pi} \left(\int_0^b e^{-r^2} r dr\right) d\theta$$
$$= \lim_{b \to +\infty} \int_0^{2\pi} \left[-\frac{1}{2} e^{-r^2}\right]_0^b d\theta$$
$$= \lim_{b \to +\infty} \frac{1}{2} \int_0^{2\pi} (1 - e^{-b^2}) d\theta$$
$$= \lim_{b \to +\infty} \pi(1 - e^{-b^2}) = \pi.$$

所以

$$I = \frac{\sqrt{V}}{2} = \frac{\sqrt{\pi}}{2}.$$

例5 求二重积分

$$I = \iint_S (2 - x^2 - y^2) dA,$$

这里 $S = \{(x,y) \mid x^2 + y^2 \leqslant 1\}$.

解 根据 S 的对称性,设 S_1 为 S 在第一象限的部分,有

$$I = \iint_S (2 - x^2 - y^2) dA$$

$$= 4\iint_{S_1}(2-x^2-y^2)\,\mathrm{d}A.$$

In polar coordinates, there is

$$S_1 = \left\{(r,\theta)\,\bigg|\,0\leqslant r\leqslant 1, 0\leqslant\theta\leqslant\frac{\pi}{2}\right\},$$

and $2-x^2-y^2 = 2-r^2$, so

$$I = 4\int_0^{\frac{\pi}{2}}\int_0^1 (2-r^2)r\,\mathrm{d}r\,\mathrm{d}\theta$$
$$= 3\int_0^{\frac{\pi}{2}}\mathrm{d}\theta = \frac{3}{2}\pi.$$

3.5 Applications of Double Integral

The most obvious application of double integral is calculating volumes of solids. This application of double integral has been made a detailed explanation, so now we turn to other applications.

1. The Quality of Flat Sheet

Suppose that the area occupied by a sheet in Oxy-plane is S, and the density at point (x,y) be denoted by $\delta(x,y)$. We want to solve the quality m of the sheet.

Use two straight lines which are parallel to the x-axis and y-axis respectively to divide S into n areas $R_1, R_2, \cdots, R_n$. These areas are all rectangle except for the area containing the border, as shown in Figure 3.36.

Pick a point $(\overline{x}_k, \overline{y}_k)$ in R_k, then the mass of R_k is approximately $\delta(\overline{x}_k, \overline{y}_k)A(R_k)$, and the mass of the whole sheet is

$$m \approx \sum_{k=1}^n \delta(\overline{x}_k, \overline{y}_k)A(R_k),$$

where $A(R_k)$ is the area of R_k. Let the maximum diameter of the divided small area tends to zero, and take the limit to get the quality m of the sheet, which is a double integral:

$$m = \iint_S \delta(x,y)\,\mathrm{d}A. \tag{3.12}$$

3.5 Applications of Double Integral
3.5 二重积分的应用

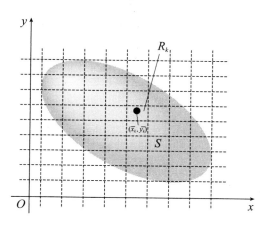

Figure 3.36
图 3.36

Example 1 The area S occupied by a sheet in Oxy-plane is surrounded by x-axis, straight line $x=8$ and curve $y=x^{2/3}$, its density is $\delta(x,y)=xy$ (Figure 3.37). Solve its quality.

例1 一块薄板在 Oxy 平面上所占的区域 S 由 x 轴、直线 $x=8$ 和曲线 $y=x^{2/3}$ 所围成，其密度为 $\delta(x,y)=xy$（图 3.37），求它的质量.

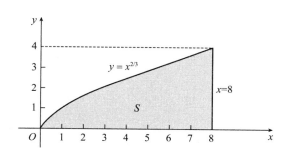

Figure 3.37
图 3.37

Solution According to the formula (3.12), the quality of the sheet is

$$m = \iint_S xy\,dA = \int_0^8 \int_0^{x^{2/3}} xy\,dy\,dx$$
$$= \int_0^8 \left[\frac{xy^2}{2}\right]_0^{x^{2/3}} dx = \frac{1}{2}\int_0^8 x^{7/3}\,dx$$
$$= \frac{1}{2}\left[\frac{3}{10}x^{10/3}\right]_0^8 = \frac{768}{5}.$$

解 根据公式(3.12)，得该薄板的质量为

$$m = \iint_S xy\,dA = \int_0^8 \int_0^{x^{2/3}} xy\,dy\,dx$$
$$= \int_0^8 \left[\frac{xy^2}{2}\right]_0^{x^{2/3}} dx = \frac{1}{2}\int_0^8 x^{7/3}\,dx$$
$$= \frac{1}{2}\left[\frac{3}{10}x^{10/3}\right]_0^8 = \frac{768}{5}.$$

2. Center of Mass of Flat Sheet

Now we can determine the center of mass of sheet by the double integral. The qualities of n particles at points (x_1, y_1), $(x_2, y_2), \cdots, (x_n, y_n)$ are $m_1, m_2, \cdots, m_n$ respectively. Then the total static moment of the particles system formed by these n particles on y-axis and x-axis respectively are

$$M_y = \sum_{k=1}^{n} x_k m_k, \quad M_x = \sum_{k=1}^{n} y_k m_k.$$

Moreover, the coordinates $(\overline{x}, \overline{y})$ of the center of mass (balance point) are

$$\overline{x} = \frac{M_y}{m} = \frac{\sum_{k=1}^{n} x_k m_k}{\sum_{k=1}^{n} m_k},$$

$$\overline{y} = \frac{M_x}{m} = \frac{\sum_{k=1}^{n} y_k m_k}{\sum_{k=1}^{n} m_k},$$

where $m = \sum_{k=1}^{n} m_k$ is the total quality of n particles.

Consider the centre of mass of the sheet covering a region S in the Oxy-plane, whose density is $\delta(x, y)$. Doing the same split on S (as shown in Figure 3.36), we get n small areas $R_1, R_2, \cdots, R_n$. Suppose the quality of the small sheet corresponding to each R_k is approximately centered at the point $(\overline{x}_k, \overline{y}_k)$, $k = 1, 2, \cdots, n$. At last, let the maximun diameter of the divided small area tends to zero. Taking its limit we could get the static moment of sheet on y-axis and x-axis respectively are

$$M_y = \iint_S x \delta(x, y) \, dA,$$

$$M_x = \iint_S y \delta(x, y) \, dA.$$

So we could have the center of mass of the flat sheet formula:

$$\overline{x} = \frac{M_y}{m} = \frac{\iint_S x \delta(x, y) \, dA}{\iint_S \delta(x, y) \, dA},$$

2. 平面薄板的质心

下面我们由二重积分来确定薄板的质心. 我们知道，如果 n 个位于点 (x_1, y_1), $(x_2, y_2), \cdots, (x_n, y_n)$ 的质点质量分别为 $m_1, m_2, \cdots, m_n$，则这 n 个质点构成的质点系关于 y 轴和 x 轴的总静矩分别为

$$M_y = \sum_{k=1}^{n} x_k m_k, \quad M_x = \sum_{k=1}^{n} y_k m_k,$$

从而质心的坐标 $(\overline{x}, \overline{y})$（平衡点）为

$$\overline{x} = \frac{M_y}{m} = \frac{\sum_{k=1}^{n} x_k m_k}{\sum_{k=1}^{n} m_k},$$

$$\overline{y} = \frac{M_x}{m} = \frac{\sum_{k=1}^{n} y_k m_k}{\sum_{k=1}^{n} m_k},$$

其中 $m = \sum_{k=1}^{n} m_k$ 为 n 个质点的总质量.

现在考虑在 Oxy 平面内所占区域为 S 的薄板的质心，薄板的密度为 $\delta(x, y)$. 同样，对 S 作如图 3.36 所示的分割，得到 n 个小区域 $R_1, R_2, \cdots, R_n$. 假设每个 R_k 对应的小块薄板的质量近似集中在点 $(\overline{x}_k, \overline{y}_k)$ 处，$k = 1, 2, \cdots, n$. 最后，令分割的小区域的最大直径趋于零，取极限，可以得到薄板关于 y 轴和 x 轴的静矩分别为

$$M_y = \iint_S x \delta(x, y) \, dA,$$

$$M_x = \iint_S y \delta(x, y) \, dA.$$

于是得到薄板的质心公式

$$\overline{x} = \frac{M_y}{m} = \frac{\iint_S x \delta(x, y) \, dA}{\iint_S \delta(x, y) \, dA},$$

$$\bar{y} = \frac{M_x}{m} = \frac{\iint\limits_S y\delta(x,y)\,dA}{\iint\limits_S \delta(x,y)\,dA},$$

where $m = \iint\limits_S \delta(x,y)\,dA$ is the quality of the sheet.

Example 2 Find the center of mass of the sheet of Example 1.

Solution In Example 1, we showed that the quality m of the sheet is $\frac{768}{5}$. The static moments M_y and M_x of the flat sheet on y-axis and x-axis are

$$M_y = \iint\limits_S x\delta(x,y)\,dA = \int_0^8 \int_0^{x^{2/3}} x^2 y\,dy\,dx$$
$$= \frac{1}{2}\int_0^8 x^{10/3}\,dx = \frac{12288}{13},$$
$$M_x = \iint\limits_S y\delta(x,y)\,dA = \int_0^8 \int_0^{x^{2/3}} xy^2\,dy\,dx$$
$$= \frac{1}{3}\int_0^8 x^3\,dx = \frac{1024}{3}.$$

So the center of mass of the flat sheet is

$$\bar{x} = \frac{M_y}{m} = \frac{80}{13}, \quad \bar{y} = \frac{M_x}{m} = \frac{20}{9}.$$

Example 3 Find the center of mass of a sheet S in the shape of a quarter disk (Figure 3.38). Suppose that its density at any point is proportional to the distance from the point to the center.

$$\bar{y} = \frac{M_x}{m} = \frac{\iint\limits_S y\delta(x,y)\,dA}{\iint\limits_S \delta(x,y)\,dA},$$

其中 $m = \iint\limits_S \delta(x,y)\,dA$ 为薄板的质量.

例 2 求例 1 中薄板的质心.

解 在例 1 中,我们给出了薄板的质量是 $\frac{768}{5}$. 薄板关于 y 轴和 x 轴的静矩 M_y 和 M_x 是

$$M_y = \iint\limits_S x\delta(x,y)\,dA = \int_0^8 \int_0^{x^{2/3}} x^2 y\,dy\,dx$$
$$= \frac{1}{2}\int_0^8 x^{10/3}\,dx = \frac{12288}{13},$$
$$M_x = \iint\limits_S y\delta(x,y)\,dA = \int_0^8 \int_0^{x^{2/3}} xy^2\,dy\,dx$$
$$= \frac{1}{3}\int_0^8 x^3\,dx = \frac{1024}{3}.$$

于是我们得到薄板的质心

$$\bar{x} = \frac{M_y}{m} = \frac{80}{13}, \quad \bar{y} = \frac{M_x}{m} = \frac{20}{9}.$$

例 3 求一块四分之一圆形薄板 S 的质心(图 3.38),假设其任一点处的密度与该点到圆的中心的距离成比例.

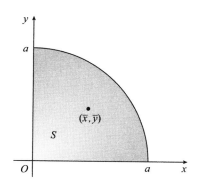

Figure 3.38
图 3.38

Solution Cartesian coordinate system shown in Figure 3.38. From the assumption that the density of the sheet is
$$\delta(x,y) = k\sqrt{x^2+y^2},$$
where k is a constant. The shape of S (as shown in Figure 3.38) is represented by polar coordinates as
$$S = \left\{(r,\theta) \,\middle|\, 0 \leqslant r \leqslant a, 0 \leqslant \theta \leqslant \frac{\pi}{2}\right\}.$$
So the quality of the sheet is
$$\begin{aligned}m &= \iint_S \delta(x,y)\,dA \\ &= \iint_S k\sqrt{x^2+y^2}\,dA \\ &= k\int_0^{\frac{\pi}{2}}\int_0^a r^2\,dr\,d\theta \\ &= k\int_0^{\frac{\pi}{2}} \frac{a^3}{3}\,d\theta = \frac{k\pi a^3}{6}.\end{aligned}$$
The static moment on the x-axis of the sheet is
$$\begin{aligned}M_y &= \iint_S xk\sqrt{x^2+y^2}\,dA \\ &= k\int_0^{\frac{\pi}{2}}\int_0^a r\cos\theta \cdot r^2\,dr\,d\theta \\ &= k\int_0^{\frac{\pi}{2}} \frac{a^4}{4}\cos\theta\,d\theta = \frac{ka^4}{4},\end{aligned}$$
we conclude that
$$\overline{x} = \frac{M_y}{m} = \frac{\dfrac{ka^4}{4}}{\dfrac{k\pi a^3}{6}} = \frac{3a}{2\pi}.$$

Because of the symmetry with $y = x$ of the sheet, we recognize that $\overline{y} = \overline{x}$, so no further calculation is needed. Therefore, the center of mass of the sheet is $\left(\dfrac{3a}{2\pi}, \dfrac{3a}{2\pi}\right)$.

3. The Moment of Inertia of a Flat Sheet

Let the area occupied by the sheet in Oxy-plane is S, and the density of the sheet is $\delta(x,y)$. Let's consider the moment of inertia of the sheet about the coordinate axes.

If we split it the same as Figure 3.36 and suppose the quality of each R_k, is approximately centered at point $(\overline{x}, \overline{y})$, then the moment inertia of each R_k with respect to x-axis and

y-axis, respectively, are approximated as
$$y^2\delta(\overline{x},\overline{y})A(R_k), \quad x^2\delta(\overline{x},\overline{y})A(R_k).$$
Add together, then take the limit, we can get the formula of the moments of inertia of the sheet I_x and I_y about the x-axis,
$$I_x = \iint_S y^2\delta(x,y)\mathrm{d}A,$$
$$I_y = \iint_S x^2\delta(x,y)\mathrm{d}A.$$

Example 4 Find the moments of inertia about x-axis and y-axis of the sheet of Example 1.

Solution Where $\delta(x,y)=xy$, so the moment of inertia of sheet about the x-axis and y-axis respectively are
$$I_x = \iint_S xy^3\mathrm{d}A = \int_0^8\int_0^{x^{2/3}} xy^3\mathrm{d}y\mathrm{d}x$$
$$= \frac{1}{4}\int_0^8 x^{11/3}\,\mathrm{d}x = \frac{6144}{7},$$
$$I_y = \iint_S x^3 y\mathrm{d}A = \int_0^8\int_0^{x^{2/3}} x^3 y\mathrm{d}y\mathrm{d}x$$
$$= \frac{1}{2}\int_0^8 x^{13/3}\,\mathrm{d}x = 6144.$$

Exercises 3
习题 3

1. Let $R = \{(x,y): 1\leqslant x\leqslant 4, 0\leqslant y\leqslant 2\}$. Calculate $\iint_R f(x,y)\mathrm{d}A$, where $f(x,y)$ is given by

(1) $f(x,y)=\begin{cases}2, & 1\leqslant x<3, 0\leqslant y\leqslant 2,\\ 3, & 3\leqslant x\leqslant 4, 0\leqslant y\leqslant 2;\end{cases}$

(2) $f(x,y)=\begin{cases}-1, & 1\leqslant x\leqslant 4, 0\leqslant y<1,\\ 2, & 1\leqslant x\leqslant 4, 1\leqslant y\leqslant 2;\end{cases}$

(3) $f(x,y)=\begin{cases}2, & 1\leqslant x<3, 0\leqslant y<1,\\ 1, & 1\leqslant x<3, 1\leqslant y\leqslant 2,\\ 3, & 3\leqslant x\leqslant 4, 0\leqslant y\leqslant 2.\end{cases}$

2. Sketch the graph of the double integral of the following function over the rectangular area R:
$$R=\{(x,y)|0\leqslant x\leqslant 2, 0\leqslant y\leqslant 3\}.$$
(1) $\iint\limits_R 3\mathrm{d}A$;

(2) $\iint\limits_R (x+1)\mathrm{d}A$;

(3) $\iint\limits_R (y+1)\mathrm{d}A$;

(4) $\iint\limits_R (x-y+4)\mathrm{d}A$.

3. Calculate $\iint\limits_R (6-y)\mathrm{d}A$, where
$$R=\{(x,y)|0\leqslant x\leqslant 1, 0\leqslant y\leqslant 1\}.$$
This double integral represent the volume of a certain solid. Sketch this solid.

4. Calculate $\iint\limits_R (1+x)\mathrm{d}A$, where
$$R=\{(x,y)|0\leqslant x\leqslant 2, 0\leqslant y\leqslant 1\}.$$

5. Use the comparison property of double integral to show that if $f(x,y)\geqslant 0$ on R, then
$$\iint\limits_R f(x,y)\mathrm{d}A \geqslant 0.$$

6. Calculate the following iterated integrals:

(1) $\int_0^2\int_0^3 (9-x)\mathrm{d}y\mathrm{d}x$;

(2) $\int_{-2}^2\int_0^1 (9-x^2)\mathrm{d}y\mathrm{d}x$;

(3) $\int_0^2\int_1^3 x^2 y\mathrm{d}y\mathrm{d}x$;

(4) $\int_{-1}^4\int_1^2 (x+y^2)\mathrm{d}y\mathrm{d}x$.

7. Calculate the following double integrals:

(1) $\iint\limits_R xy^3\mathrm{d}A$, where
$$R=\{(x,y)|0\leqslant x\leqslant 1, -1\leqslant y\leqslant 1\};$$

(2) $\iint\limits_R \sin(x+y)\mathrm{d}A$, where
$$R=\left\{(x,y)\Big|0\leqslant x\leqslant \frac{\pi}{2}, 0\leqslant y\leqslant \frac{\pi}{2}\right\}.$$

2. 画出下列函数在矩形区域
$$R=\{(x,y)|0\leqslant x\leqslant 2, 0\leqslant y\leqslant 3\}$$
上的二重积分的图形：

(1) $\iint\limits_R 3\mathrm{d}A$;

(2) $\iint\limits_R (x+1)\mathrm{d}A$;

(3) $\iint\limits_R (y+1)\mathrm{d}A$;

(4) $\iint\limits_R (x-y+4)\mathrm{d}A$.

3. 计算 $\iint\limits_R (6-y)\mathrm{d}A$，这里
$$R=\{(x,y)|0\leqslant x\leqslant 1, 0\leqslant y\leqslant 1\}.$$
这个二重积分表示某个立体的体积，描绘出这个立体.

4. 计算 $\iint\limits_R (1+x)\mathrm{d}A$，这里
$$R=\{(x,y)|0\leqslant x\leqslant 2, 0\leqslant y\leqslant 1\}.$$

5. 运用二重积分的大小比较性质来证明：若 $f(x,y)\geqslant 0$ 在 R 上成立，则
$$\iint\limits_R f(x,y)\mathrm{d}A \geqslant 0.$$

6. 计算下列累次积分：

(1) $\int_0^2\int_0^3 (9-x)\mathrm{d}y\mathrm{d}x$;

(2) $\int_{-2}^2\int_0^1 (9-x^2)\mathrm{d}y\mathrm{d}x$;

(3) $\int_0^2\int_1^3 x^2 y\mathrm{d}y\mathrm{d}x$;

(4) $\int_{-1}^4\int_1^2 (x+y^2)\mathrm{d}y\mathrm{d}x$.

7. 计算下列二重积分：

(1) $\iint\limits_R xy^3\mathrm{d}A$，其中
$$R=\{(x,y)|0\leqslant x\leqslant 1, -1\leqslant y\leqslant 1\};$$

(2) $\iint\limits_R \sin(x+y)\mathrm{d}A$，其中
$$R=\left\{(x,y)\Big|0\leqslant x\leqslant \frac{\pi}{2}, 0\leqslant y\leqslant \frac{\pi}{2}\right\}.$$

8. Calculate $\int_0^1\int_0^1 xy e^{x^2+y^2} \mathrm{d}y\mathrm{d}x$.

9. Calculate $\int_0^{\sqrt{3}}\int_0^1 \dfrac{8x}{(x^2+y^2+1)^2}\mathrm{d}y\mathrm{d}x$.

Hint: Exchange the order of integration.

10. Calculate the following iterated integrals:

(1) $\int_0^1\int_0^{3x} x^2 \mathrm{d}y\mathrm{d}x$;

(2) $\int_1^2\int_0^{x-1} y\mathrm{d}y\mathrm{d}x$;

(3) $\int_{-1}^3\int_0^{3y}(x^2+y^2)\mathrm{d}x\mathrm{d}y$;

(4) $\int_{-3}^1\int_0^x (x^2-y^3)\mathrm{d}y\mathrm{d}x$;

(5) $\int_1^3\int_{-y}^{2y} x e^{y^3}\mathrm{d}x\mathrm{d}y$;

(6) $\int_1^5\int_0^x \dfrac{3}{x^2+y^2}\mathrm{d}y\mathrm{d}x$.

11. Sketch the indicated solid. Then find its volume by double integral.

(1) Tetrahedron bounded by the coordinate planes and the plane $z=6-2x-3y$;

(2) Tetrahedron bounded by the coordinate planes and the plane $3x+4y+z-12=0$;

(3) Solid in the first octant bounded by the coordinate planes
$$2x+y-4=0 \quad \text{and} \quad 8x+y-4z=0.$$

12. Calculate $\iint_S \sin(xy^2)\mathrm{d}A$, where S is the annulus $\{(x,y)\mid 1\leqslant x^2+y^2\leqslant 4\}$.

Hint: Use symmetry.

13. Calculate $\iint_S \sin y^3 \mathrm{d}A$, where S is the closed region bounded by curve $y=\sqrt{x}$, line $y=2$ and $x=0$.

14. Calculate $\iint_S x^2 \mathrm{d}A$, where S is the closed region between the ellipse $x^2+2y^2=4$ and the circle $x^2+y^2=4$.

15. Calculate the following double integrals in polar coordinates:

(1) $\iint\limits_S r^2 \sin\theta \, dA$, where
$S = \left\{ (r,\theta) \,\middle|\, 0 \leqslant r \leqslant \cos\theta, 0 \leqslant \theta \leqslant \dfrac{\pi}{2} \right\}$;

(2) $\iint\limits_S r \, dA$, where
$S = \left\{ (r,\theta) \,\middle|\, 0 \leqslant r \leqslant \sin\theta, 0 \leqslant \theta \leqslant \dfrac{\pi}{2} \right\}$;

(3) $\iint\limits_S r^2 \, dA$, where
$S = \{ (r,\theta) \,|\, 0 \leqslant r \leqslant \sin\theta, 0 \leqslant \theta \leqslant \pi \}$;

(4) $\iint\limits_S r \sin\theta \, dA$, where
$S = \{ (r,\theta) \,|\, 0 \leqslant r \leqslant 1 - \cos\theta, 0 \leqslant \theta \leqslant \pi \}$;

(5) $\iint\limits_S r \sin\dfrac{\theta}{4} \, dA$, where
$S = \{ (r,\theta) \,|\, 0 \leqslant r \leqslant 2, 0 \leqslant \theta \leqslant \pi \}$;

(6) $\iint\limits_S r \, dA$, where
$S = \{ (r,\theta) \,|\, 0 \leqslant r \leqslant \theta, 0 \leqslant \theta \leqslant 2\pi \}$;

(7) $\iint\limits_S 2e^{x^2+y^2} \, dA$, where S is the closed region that is surrounded by the circumference $x^2 + y^2 = 4$;

(8) $\iint\limits_S \ln(1 + x^2 + y^2) \, dA$ where S is the closed region in the first quadrant that is surrounded by the circumference $x^2 + y^2 = 1$ and coordinate axes;

(9) $\iint\limits_S \arctan\dfrac{y}{x} \, dA$, where S is the closed region in the first quadrant that is surrounded by the circumferences $x^2 + y^2 = 4$, $x^2 + y^2 = 1$, and the lines $y = 0, y = x$.

16. Sketch the integral areas of the following double integrals and calculate the double integrals:

(1) $\iint\limits_S x\sqrt{y} \, dA$, where S is the closed region surrounded by two parabolas $y = x^2$ and $x = y^2$;

(2) $\iint\limits_S (x^2 + y^2 - x) \, dA$, where S is the closed region surrounded by the lines $y = 2x, y = x$ and $y = 2$;

(3) $\iint\limits_{S} \dfrac{\sin y}{y} dA$, where S is the closed region surrounded by two parabolas $y=x^2$ and $y=x$.

17. Choose the appropriate coordinates to calculate the following double integrals:

(1) $\iint\limits_{S} \dfrac{x^2}{y^2} dA$, where S is the closed region surrounded by the lines $x=2$, $y=x$ and curve $xy=1$;

(2) $\iint\limits_{S} \sqrt{\dfrac{1-x^2-y^2}{1+x^2+y^2}} dA$, where S is the closed region in the first quadrant that is surrounded by the circumference $x^2+y^2=1$ and coordinate axes;

(3) $\iint\limits_{S} (x^2+y^2) dA$, where S is the closed region surrounded by the lines $y=x+a$, $y=x$, $y=a$ and $y=3a$ ($a>0$);

(4) $\iint\limits_{S} \sqrt{x^2+y^2} dA$, where S is the circular closed region $\{(x,y) \mid a^2 \leqslant x^2+y^2 \leqslant b^2\}$ ($0<a<b$).

Chapter 4　Infinite Series
第 4 章　无穷级数

Generally speaking, an infinite sequence
$$a_1, a_2, a_3, a_4, \cdots \qquad (4.1)$$
is an ordered arrangement of real numbers, one for each corresponds to a positive integer. More formally, an **infinite sequence** is a function whose domain is the set of positive integers and whose range is a set of real numbers. Infinite sequence is also called **sequence** for short. We may denote a sequence (4.1) by $\{a_n\}_{n=1}^{\infty}$ or $\{a_n\}$. Occasionally, we will extend the notion slightly by allowing the domain to consist of all integers greater than or equal to a specific integer, as in
$$a_0, a_1, a_2, \cdots \quad \text{and} \quad a_5, a_6, a_7, \cdots$$
which are also written as $\{a_n\}_{n=0}^{\infty}$ and $\{a_n\}_{n=5}^{\infty}$, respectively.

A sequence may be specified by giving enough initial terms to establish a pattern, as in
$$3, 5, 7, 9, 11, \cdots$$
by an explicit formula for the nth term, as in
$$a_n = 2n+1 \quad (n \geqslant 1)$$
or by a recursion formula
$$a_1 = 3, \quad a_n = a_{n-1} + 2 \quad (n \geqslant 2).$$
Note that each of our three illustrations describes the same sequence.

Here are four sequences given by explicit formulas and the first few terms:

(1) $a_n = \dfrac{n+1}{n}$ $(n \geqslant 1)$:
$$2, \frac{3}{2}, \frac{4}{3}, \frac{5}{4}, \cdots;$$

(2) $b_n = \dfrac{(-1)^n}{n}$ $(n \geqslant 1)$:
$$-1, \frac{1}{2}, -\frac{1}{3}, \frac{1}{4}, \cdots;$$

一般来说，一个无穷数列
$$a_1, a_2, a_3, a_4, \cdots \qquad (4.1)$$
是一组有序排列的实数，每个都对应一个正整数. 更确切地说，一个**无穷数列**是一个定义域为正整数集，值域为实数集的函数. 无穷数列也简称**数列**. 我们可以用 $\{a_n\}_{n=1}^{\infty}$ 或 $\{a_n\}$ 来表示数列(4.1). 有时，我们也通过定义域由大于或等于某个特殊整数组成来稍微扩展数列的表示方法，例如将数列
$$a_0, a_1, a_2, \cdots \quad \text{和} \quad a_5, a_6, a_7, \cdots$$
分别记作 $\{a_n\}_{n=0}^{\infty}$ 和 $\{a_n\}_{n=5}^{\infty}$.

一个数列可以通过给定足够的初始条件建立模型来表示. 例如，数列
$$3, 5, 7, 9, 11, \cdots$$
可通过第 n 项的显示公式表示为
$$a_n = 2n+1 \quad (n \geqslant 1),$$
或由一个递推公式来表示：
$$a_1 = 3, \quad a_n = a_{n-1} + 2 \quad (n \geqslant 2).$$
注意，我们给出三种表示法描述的都是同一个数列.

下面是四个由显示公式给出的数列及它们各自最初的几项：

(1) $a_n = \dfrac{n+1}{n}$ $(n \geqslant 1)$:
$$2, \frac{3}{2}, \frac{4}{3}, \frac{5}{4}, \cdots;$$

(2) $b_n = \dfrac{(-1)^n}{n}$ $(n \geqslant 1)$:
$$-1, \frac{1}{2}, -\frac{1}{3}, \frac{1}{4}, \cdots;$$

(3) $c_n = 1 + \left(-\dfrac{1}{n}\right)^n$ $(n \geq 1)$：

$$0, \frac{5}{4}, \frac{26}{27}, \frac{255}{256}, \cdots;$$

(4) $d_n = 2$ $(n \geq 1)$：

$$2, 2, 2, 2, \cdots.$$

Generally, for the given sequence

$$a_1, a_2, \cdots, a_n, \cdots,$$

the expression formed by the sequence

$$a_1 + a_2 + \cdots + a_n + \cdots$$

is called **the infinite series of the constant term**, referred to the **infinite series** or the **series**, written as $\sum\limits_{n=1}^{\infty} a_n$, that is

$$\sum_{n=1}^{\infty} a_n = a_1 + a_2 + a_3 + \cdots + a_n + \cdots,$$

where the nth term is said to be the **general term** of the series, moreover, the nth **partial sum** of the series (referred to the **partial sum**) is S_n, given by

$$S_n = a_1 + a_2 + \cdots + a_n.$$

(3) $c_n = 1 + \left(-\dfrac{1}{n}\right)^n$ $(n \geq 1)$：

$$0, \frac{5}{4}, \frac{26}{27}, \frac{255}{256}, \cdots;$$

(4) $d_n = 2$ $(n \geq 1)$：

$$2, 2, 2, 2, \cdots.$$

通常，对于给定的数列

$$a_1, a_2, \cdots, a_n, \cdots,$$

由该数列形成的表达式

$$a_1 + a_2 + \cdots + a_n + \cdots$$

称为**常数项无穷级数**，简称**无穷级数**或**级数**，记作 $\sum\limits_{n=1}^{\infty} a_n$，即

$$\sum_{n=1}^{\infty} a_n = a_1 + a_2 + a_3 + \cdots + a_n + \cdots,$$

这里第 n 项称为这个级数的**通项**. 此外，该级数的**前 n 项部分和**（简称**部分和**）记为 S_n，即

$$S_n = a_1 + a_2 + \cdots + a_n.$$

4.1 Determine Whether the Infinite Series Converges or Diverges
4.1 判断无穷级数的敛散性

1. The Concept of Convergence and Divergence of Series

While it's possible to add two numbers, three numbers, one hundred numbers, or even one million numbers, but it's impossible to add an infinite number of numbers. To form an infinite series we begin with an infinite sequence

$$a_1, a_2, a_3, \cdots.$$

We can't form the sum of all the $a_n (n=1, 2, \cdots)$ (there is an infinite number of the term), but we can form the partial sums

$$S_1 = a_1 = \sum_{k=1}^{1} a_k,$$

$$S_2 = a_1 + a_2 = \sum_{k=1}^{2} a_k,$$

$$S_3 = a_1 + a_2 + a_3 = \sum_{k=1}^{3} a_k,$$

1. 级数敛散性的概念

虽然可以将两个数、三个数、一百个数甚至一百万个数相加，但不可能将无穷多个数相加. 要形成一个无穷级数，我们从一个无穷数列开始：

$$a_1, a_2, a_3, \cdots.$$

我们不能得出所有项 $a_n (n=1, 2, \cdots)$ 的和（这里有无穷多项），但是可以得到部分和

$$S_1 = a_1 = \sum_{k=1}^{1} a_k,$$

$$S_2 = a_1 + a_2 = \sum_{k=1}^{2} a_k,$$

$$S_3 = a_1 + a_2 + a_3 = \sum_{k=1}^{3} a_k,$$

……

……
$$S_n = a_1 + a_2 + \cdots + a_n = \sum_{k=1}^{n} a_k,$$
……

Definition 4.1 If the sequence $\{S_n\}$ of partial sums has the limit L. Then we say the series $\sum_{n=1}^{\infty} a_n$ **converges** to L, and L is the **sum** of the series. We write
$$L = \lim_{n \to \infty} \sum_{k=1}^{n} a_k, \quad \text{or} \quad L = \sum_{n=1}^{\infty} a_n.$$
If the limit of the sequence $\{S_n\}$ of partial sum don't exists, we say that the series $\sum_{n=1}^{\infty} a_n$ **diverges**.

Remark It is important to note that the sum of a series is not a sum in the general sense, it is a limit.

Definition 4.2 A series of the form
$$\sum_{n=1}^{\infty} ar^{n-1} = a + ar + ar^2 + ar^3 + \cdots$$
$$(a \neq 0)$$
is called a **geometric series**, where r is called the **common ratio** of the series.

Example 1 Prove the following propositions:

(1) If $|x| < 1$, then $\sum_{n=0}^{\infty} x^n$ converges, and
$$\sum_{n=0}^{\infty} x^n = \frac{1}{1-x};$$

(2) If $|x| \geqslant 1$, then $\sum_{n=0}^{\infty} x^n$ diverges.

Proof The nth partial sum of the geometric series $\sum_{n=0}^{\infty} x^n$ is
$$S_n = 1 + x + x^2 + \cdots + x^{n-1}. \quad (4.2)$$
Multiplying by x, we have
$$xS_n = x(1 + x + x^2 + \cdots + x^{n-1})$$
$$= x + x^2 + x^3 + \cdots + x^{n-1} + x^n. \quad (4.3)$$
Subtracting the equation (4.2) from the first (4.3), we find that
$$(1-x)S_n = 1 - x^n.$$
For $x \neq 1$, this gives

……
$$S_n = a_1 + a_2 + \cdots + a_n = \sum_{k=1}^{n} a_k,$$
……

定义 4.1 如果部分和数列 $\{S_n\}$ 有极限 L，则称级数 $\sum_{n=1}^{\infty} a_n$ **收敛**于 L，并称 L 为该级数的和，记作
$$L = \lim_{n \to \infty} \sum_{k=1}^{n} a_k \quad \text{或} \quad L = \sum_{n=1}^{\infty} a_n.$$
如果部分和数列 $\{S_n\}$ 的极限不存在，则称级数 $\sum_{n=1}^{\infty} a_n$ **发散**.

注 一个级数的和不是一般意义上的和，而是一个极限.

定义 4.2 形如
$$\sum_{n=1}^{\infty} ar^{n-1} = a + ar + ar^2 + ar^3 + \cdots$$
$$(a \neq 0)$$
的级数称为**几何级数**，其中 r 称为该级数的公比.

例 1 证明下列命题：

(1) 如果 $|x| < 1$，那么 $\sum_{n=0}^{\infty} x^n$ 收敛，并且
$$\sum_{n=0}^{\infty} x^n = \frac{1}{1-x};$$

(2) 如果 $|x| \geqslant 1$，那么 $\sum_{n=0}^{\infty} x^n$ 发散.

证明 几何级数 $\sum_{n=0}^{\infty} x^n$ 的前 n 项部分和是
$$S_n = 1 + x + x^2 + \cdots + x^{n-1}. \quad (4.2)$$
上式两端乘以 x，得到
$$xS_n = x(1 + x + x^2 + \cdots + x^{n-1})$$
$$= x + x^2 + x^3 + \cdots + x^{n-1} + x^n. \quad (4.3)$$
由(4.2)式减去(4.3)式，可得
$$(1-x)S_n = 1 - x^n.$$
当 $x \neq 1$ 时，得到

$$S_n = \frac{1-x^n}{1-x}. \qquad (4.4)$$

If $|x|<1$, then $x^n \to 0$, and the right of the equation (4.4) become $\frac{1}{1-x}$, then

$$\lim_{n\to 0} S_n = \lim_{n\to 0} \frac{1-x^n}{1-x} = \frac{1}{1-x}.$$

This proves (1).

Now let us prove (2). For $x=1$, we use equation (4.2) and device that $S_n = n$. Obviously, $\lim_{n\to\infty} S_n = +\infty$, $\sum_{n=0}^{\infty} x_n$ diverges. For $x=-1$, we use formula (4.2) and deduce that:

If n is odd, then $S_n = 1$;

If n is even, then $S_n = 0$.

For the sequence $\{S_n\}$ of partial sum likes this $1,0,1,0,1,0,\cdots$, the limit of sequence $\{S_n\}$ of partial sum does not exist. By Definition 4.1, we have the series $\sum_{k=0}^{\infty} x^k$ diverges at $x=-1$.

For $x \neq 1$ with $|x|>1$, we use formula (4.4). Since in this instance, we have

$$\lim_{n\to\infty} S_n = \lim_{n\to\infty} \frac{1-x^n}{1-x} = -\infty.$$

The limit of sequence $\{S_n\}$ of partial sum does not exist, hence the series $\sum_{k=0}^{\infty} x^k$ diverges.

The geometric series appears more than once in this book, we should pay special attention. It is one of the few series where we can actually give an explicit formula for S_n.

Example 2 Determine whether or not the following series converges:

$$\sum_{n=0}^{\infty} \frac{1}{(n+1)(n+2)}.$$

Solution In order to determine whether or not this series converges, we must examine the partial sum. Since

$$\frac{1}{(n+1)(n+2)} = \frac{1}{n+1} - \frac{1}{n+2},$$

we use partial fraction decomposition to write

$$S_n = \frac{1}{1\times 2} + \frac{1}{2\times 3} + \cdots + \frac{1}{n(n+1)}$$
$$+ \frac{1}{(n+1)(n+2)}$$
$$= \left(1 - \frac{1}{2}\right) + \left(\frac{1}{2} - \frac{1}{3}\right) + \cdots$$
$$+ \left(\frac{1}{n} - \frac{1}{n+1}\right) + \left(\frac{1}{n+1} - \frac{1}{n+2}\right)$$
$$= 1 - \frac{1}{2} + \frac{1}{2} - \frac{1}{3} + \cdots + \frac{1}{n} - \frac{1}{n+1}$$
$$+ \frac{1}{n+1} - \frac{1}{n+2}.$$

Since all but the first and last occur in pairs with opposite signs, the sum collapses to give
$$S_n = 1 - \frac{1}{n+2}.$$

Obviously, as $n \to \infty$, $S_n \to 1$. This means that the series converges to 1, that is
$$\lim_{n\to\infty} S_n = \lim_{n\to\infty}\left(1 - \frac{1}{n+2}\right) = 1.$$

Therefore,
$$\sum_{n=0}^{\infty} \frac{1}{(n+1)(n+2)} = 1.$$

2. The Basic Property of the Series

Property 1 (The Necessary Condition for the Convergence of Series) If the series $\sum_{n=1}^{\infty} a_n$ converges, then
$$\lim_{n\to\infty} a_n = 0.$$
Equivalently, if $\lim_{n\to\infty} a_n \neq 0$ or if $\lim_{n\to\infty} a_n$ does not exist, then the series diverges.

Proof If $\sum_{n=1}^{\infty} a_n$ converges, according to the definition of series convergence, there is a limit L of the sequence $\{S_n\}$ of partial sums exists. Namely,
$$\lim_{n\to\infty} S_n = L.$$
Obviously,
$$\lim_{n\to\infty} S_{n-1} = L.$$
Since

2. 级数的基本性质

性质 1(级数收敛的必要条件) 如果级数 $\sum_{n=1}^{\infty} a_n$ 收敛,那么
$$\lim_{n\to\infty} a_n = 0.$$
等价地,如果 $\lim_{n\to\infty} a_n \neq 0$ 或 $\lim_{n\to\infty} a_n$ 不存在,那么这个级数发散.

证明 如果 $\sum_{n=1}^{\infty} a_n$ 收敛,根据级数收敛的定义,存在部分和数列 $\{S_n\}$ 的极限 L,即
$$\lim_{n\to\infty} S_n = L.$$
显然,又有
$$\lim_{n\to\infty} S_{n-1} = L.$$
因为

$$S_n = \sum_{k=1}^{n} a_k, \quad S_{n-1} = \sum_{k=1}^{n-1} a_k,$$

that is

$$a_n = S_n - S_{n-1},$$

we have

$$\lim_{n\to\infty} a_n = \lim_{n\to\infty}(S_n - S_{n-1})$$
$$= \lim_{n\to\infty} S_n - \lim_{n\to\infty} S_{n-1}$$
$$= L - L = 0.$$

The latter part of this property is the inverse proposition of the preceding part, so it is obviously equivalent.

Property 1 does not say that if $\lim\limits_{n\to\infty} a_n = 0$, then $\sum\limits_{n=1}^{\infty} a_n$ converges. In fact, there are divergent series for which $\lim\limits_{n\to\infty} a_n = 0$. For example, the series

$$\sum_{n=1}^{\infty} \frac{1}{\sqrt{n}} = \frac{1}{\sqrt{1}} + \frac{1}{\sqrt{2}} + \cdots + \frac{1}{\sqrt{n}} + \cdots.$$

Since the partial sum of this is

$$S_n = \frac{1}{\sqrt{1}} + \frac{1}{\sqrt{2}} + \cdots + \frac{1}{\sqrt{n}} > \frac{n}{\sqrt{n}} = \sqrt{n},$$

so

$$\lim_{n\to\infty} S_n = \lim_{n\to\infty} \sqrt{n} = +\infty.$$

Therefore the series diverges. But

$$\lim_{n\to\infty} a_n = \lim_{n\to\infty} \frac{1}{\sqrt{n}} = 0.$$

Example 3 Determine whether or not the series

$$\sum_{n=0}^{\infty} \frac{n}{n+1} = 0 + \frac{1}{2} + \frac{2}{3} + \frac{3}{4} + \frac{4}{5} + \cdots$$

converges.

Solution Since

$$\lim_{n\to\infty} a_n = \lim_{n\to\infty} \frac{n}{n+1}$$
$$= \lim_{n\to\infty} \frac{1}{1 + \frac{1}{n}}$$
$$= 1 \neq 0,$$

this series diverges by property 1.

Property 2 If the series $\sum\limits_{n=1}^{\infty} a_n$ and $\sum\limits_{n=1}^{\infty} b_n$ converges, then

(1) $\sum_{n=1}^{\infty}(a_n \pm b_n)$ also converges, and
$$\sum_{n=1}^{\infty}(a_n \pm b_n) = \sum_{n=1}^{\infty}a_n \pm \sum_{n=1}^{\infty}b_n;$$

(2) if c is a constant, then $\sum_{n=1}^{\infty}ca_n$ also converges. Moreover,
$$\sum_{n=1}^{\infty}ca_n = c\sum_{n=1}^{\infty}a_n.$$

Property 3 Let c be a nonzero constant, the series $\sum_{n=1}^{\infty}a_n$ and $\sum_{n=1}^{\infty}ca_n$ convergence or divergence at the same time.

Property 4 Suppose that the series $\sum_{n=1}^{\infty}a_n$ and $\sum_{n=1}^{\infty}c_n$ both converge to L and that
$$a_n \leqslant b_n \leqslant c_n$$
for $n \in \mathbf{N}_+$, then $\sum_{n=1}^{\infty}b_n$ also converges to L.

Property 5 We can add a finite number of terms to a series or delete a finite number of terms without altering the series convergence or divergence, although in the case of convergence this usually change the sum.

4.2 The Positive Terms Series

For the series $\sum_{n=1}^{\infty}a_n$, if $a_n \geqslant 0$ for any n, the series is called a **positive terms series**.

Theorem 4.1 A positive terms series converges if and only if the sequence of partial sums is bounded above.

Remark Theorem 4.1 give us a very useful criterion to determine when positive terms series converges.

The convergence or divergence of a positive terms series is usually deduced by comparison with a series of known behavior.

Theorem 4.2 Let $\sum_{n=1}^{\infty} a_n$ and $\sum_{n=1}^{\infty} b_n$ be two positive terms series. Assume that there is a constant $c>0$ and a positive integer n_0, when $n>n_0$, we have $a_n \leqslant cb_n$ and the series $\sum_{n=1}^{\infty} b_n$ converges, then the series $\sum_{n=1}^{\infty} a_n$ converges.

Theorem 4.2 has an analogue to show that a series does not converge

Theorem 4.3 Let $\sum_{n=1}^{\infty} a_n$ and $\sum_{n=1}^{\infty} b_n$ be two positive terms series. Assume that there is a constant $c>0$ and a positive integer n_0, when $n>n_0$, we have $a_n \geqslant cb_n$ and the series $\sum_{n=1}^{\infty} b_n$ does not converge, then the series $\sum_{n=1}^{\infty} a_n$ diverges.

Example 1 Prove that the series $\sum_{n=1}^{\infty} \frac{1}{n^2}$ converges.

Proof Let us look at the series
$$\frac{1}{1^2} + \frac{1}{2^2} + \frac{1}{3^2} + \frac{1}{4^2} + \frac{1}{5^2} + \cdots + \frac{1}{7^2}$$
$$+ \frac{1}{8^2} + \cdots + \frac{1}{15^2} + \frac{1}{16^2} + \cdots.$$

We look at the groups of terms as indicated. If we decrease the denominator in each term, then we increase the fraction. We replace 3 by 2, then 5, 6, 7 by 4, and then we replace the numbers from 9 to 15 by 8, and so forth. Our partial sums of the series are therefore less than or equal to
$$\frac{1}{1^2} + \frac{1}{2^2} + \frac{1}{2^2} + \frac{1}{4^2} + \frac{1}{4^2} + \cdots + \frac{1}{4^2}$$
$$+ \frac{1}{8^2} + \cdots + \frac{1}{8^2} + \cdots \quad (\text{taking } n \text{ terms}).$$

We note that 2 occurs twice, 4 occurs four times, 8 occurs eight times, and so forth. Our partial sums of the series are therefore less than or equal to
$$\frac{1}{1^2} + \frac{2}{2^2} + \frac{4}{4^2} + \frac{8}{8^2} + \cdots \quad (\text{taking } n \text{ terms})$$
$$= 1 + \frac{1}{2} + \frac{1}{4} + \frac{1}{8} + \cdots + \frac{1}{2^{n-1}}.$$

定理 4.2 设 $\sum_{n=1}^{\infty} a_n$ 和 $\sum_{n=1}^{\infty} b_n$ 为两个正项级数. 假设存在常数 $c>0$ 和正整数 n_0, 当 $n>n_0$ 时, 有 $a_n \leqslant cb_n$, 并且级数 $\sum_{n=1}^{\infty} b_n$ 是收敛的, 则级数 $\sum_{n=1}^{\infty} a_n$ 也是收敛的.

定理 4.2 有一个类似的版本说明一个级数不收敛的情况.

定理 4.3 设 $\sum_{n=1}^{\infty} a_n$ 和 $\sum_{n=1}^{\infty} b_n$ 为两个正项级数. 假设存在常数 $c>0$ 和正整数 n_0, 当 $n>n_0$ 时, 有 $a_n \geqslant cb_n$, 且级数 $\sum_{n=1}^{\infty} b_n$ 发散, 那么级数 $\sum_{n=1}^{\infty} a_n$ 也是发散的.

例 1 证明: 级数 $\sum_{n=1}^{\infty} \frac{1}{n^2}$ 收敛.

证明 让我们来看这个级数
$$\frac{1}{1^2} + \frac{1}{2^2} + \frac{1}{3^2} + \frac{1}{4^2} + \frac{1}{5^2} + \cdots + \frac{1}{7^2}$$
$$+ \frac{1}{8^2} + \cdots + \frac{1}{15^2} + \frac{1}{16^2} + \cdots.$$

如上式所示, 我们看每一项的组成. 如果我们在每项中减少分母, 则增加了分数. 我们用 2 代替 3, 再用 4 代替 5,6,7, 然后用 8 代替从 9 到 15 的数, 以此类推. 因此, 该级数的部分和小于或等于
$$\frac{1}{1^2} + \frac{1}{2^2} + \frac{1}{2^2} + \frac{1}{4^2} + \frac{1}{4^2} + \cdots + \frac{1}{4^2}$$
$$+ \frac{1}{8^2} + \cdots + \frac{1}{8^2} + \cdots \quad （取 n 项）.$$

我们注意到 2 出现 2 次, 4 出现 4 次, 8 出现 8 次, 以此类推. 因此, 该级数的部分和小于或等于
$$\frac{1}{1^2} + \frac{2}{2^2} + \frac{4}{4^2} + \frac{8}{8^2} + \cdots \quad （取 n 项）$$
$$= 1 + \frac{1}{2} + \frac{1}{4} + \frac{1}{8} + \cdots + \frac{1}{2^{n-1}},$$

Thus the partial sums of the series are less than or equal to the partial sums of the geometric series $\sum_{n=1}^{\infty} \frac{1}{2^{n-1}}$. This geometric series convergence, thus its partial sums series are bounded. So the partial sums series of $\sum_{n=1}^{\infty} \frac{1}{n^2}$ are bounded. Hence our series $\sum_{n=1}^{\infty} \frac{1}{n^2}$ converges.

Generally we have the following result:

The series
$$\sum_{n=1}^{\infty} \frac{1}{n^p} = 1 + \frac{1}{2^p} + \frac{1}{3^p} + \cdots + \frac{1}{n^p} + \cdots$$
is called a *p*-**series**, where p is a constant.

Proposition 1 If $p > 1$, the *p*-series converges; and if $p \leqslant 1$, then the *p*-series diverges.

In particular, when $p = 1$, the *p*-series is
$$\sum_{n=1}^{\infty} \frac{1}{n} = 1 + \frac{1}{2} + \frac{1}{3} + \cdots + \frac{1}{n} + \cdots,$$
called a **harmonic series**, and it is divergent.

Example 2 Determine whether the series $\sum_{n=1}^{\infty} \frac{n^2}{n^3+1}$ converges.

Solution We write $\frac{n^2}{n^3+1}$ as
$$\frac{n^2}{n^3+1} = \frac{1}{n + \frac{1}{n^2}} = \frac{1}{n} \cdot \frac{1}{1 + \frac{1}{n^3}}.$$
Then we see that
$$\frac{n^2}{n^3+1} \geqslant \frac{1}{2n} = \frac{1}{2} \cdot \frac{1}{n}.$$
Since the harmonic series $\sum_{n=1}^{\infty} \frac{1}{n}$ is divergent, then the series $\sum_{n=1}^{\infty} \frac{1}{2n}$ is divergent, it follows that $\sum_{n=1}^{\infty} \frac{n^2}{n^3+1}$ is divergent.

Example 3 Prove that the series $\sum_{n=1}^{\infty} \frac{1}{\ln(n+b)}$ $(b > -1)$ diverges.

于是该级数的部分和小于或等于几何级数 $\sum_{n=1}^{\infty} \frac{1}{2^{n-1}}$ 的部分和. 而此几何级数收敛,从而其部分和数列有界,故正项级数 $\sum_{n=1}^{\infty} \frac{1}{n^2}$ 的部分和数列有界. 因此级数 $\sum_{n=1}^{\infty} \frac{1}{n^2}$ 收敛.

一般地,我们有了下面的结论:

级数
$$\sum_{n=1}^{\infty} \frac{1}{n^p} = 1 + \frac{1}{2^p} + \frac{1}{3^p} + \cdots + \frac{1}{n^p} + \cdots$$
称为 *p* 级数,其中 p 是一个常数.

命题 1 如果 $p > 1$,则 *p* 级数收敛;如果 $p \leqslant 1$,那么 *p* 级数发散.

特别地,当 $p = 1$ 时,*p* 级数为
$$\sum_{n=1}^{\infty} \frac{1}{n} = 1 + \frac{1}{2} + \frac{1}{3} + \cdots + \frac{1}{n} + \cdots,$$
称之为**调和级数**,它是发散的.

例 2 判断级数 $\sum_{n=1}^{\infty} \frac{n^2}{n^3+1}$ 是否收敛.

解 我们将 $\frac{n^2}{n^3+1}$ 写成
$$\frac{n^2}{n^3+1} = \frac{1}{n + \frac{1}{n^2}} = \frac{1}{n} \cdot \frac{1}{1 + \frac{1}{n^3}}.$$
然后,我们看到
$$\frac{n^2}{n^3+1} \geqslant \frac{1}{2n} = \frac{1}{2} \cdot \frac{1}{n}.$$
由于调和级数 $\sum_{n=1}^{\infty} \frac{1}{n}$ 发散,从而级数 $\sum_{n=1}^{\infty} \frac{1}{2n}$ 发散,所以级数 $\sum_{n=1}^{\infty} \frac{n^2}{n^3+1}$ 也发散.

例 3 证明:级数 $\sum_{n=1}^{\infty} \frac{1}{\ln(n+b)}$ $(b > -1)$ 发散.

4.2 The Positive Terms Series

Proof We know that as $n \to \infty$, $\frac{\ln n}{n} \to 0$. It follows that
$$\frac{\ln(n+b)}{n+b} \to 0 \quad (n \to \infty),$$
and thus that
$$\frac{\ln(n+b)}{n} = \frac{\ln(n+b)}{n+b} \cdot \frac{n+b}{n}$$
$$\to 0 \quad (n \to \infty).$$
Thus for n sufficiently large,
$$\ln(n+b) < n \text{ is } \frac{1}{n} < \frac{1}{\ln(n+b)}.$$
Since harmonic series $\sum_{n=1}^{\infty} \frac{1}{n}$ diverges, we can conclude that series $\sum_{n=1}^{\infty} \frac{1}{\ln(n+b)}$ diverges.

Theorem 4.4 (The Comparison Test 1) Let $\sum_{n=1}^{\infty} a_n$ and $\sum_{n=1}^{\infty} b_n$ be positive terms series. If
$$\lim_{n \to \infty} \frac{a_n}{b_n} = l,$$
where l is some positive number, then series $\sum_{n=1}^{\infty} a_n$ and $\sum_{n=1}^{\infty} b_n$ converge or diverge together.

Example 4 Determine whether the series $\sum_{n=1}^{\infty} \sin \frac{\pi}{n}$ converges or diverges.

Solution Recall that as $x \to 0$, $\frac{\sin x}{x} \to 1$. As $n \to \infty$, $\frac{\pi}{n} \to 0$ and thus
$$\frac{\sin \frac{\pi}{n}}{\frac{\pi}{n}} \to 1 \quad (n \to \infty).$$
We also know that the series $\sum_{n=1}^{\infty} \frac{\pi}{n}$ diverges (because $\sum_{n=1}^{\infty} \frac{1}{n}$ diverges), thus $\sum_{n=1}^{\infty} \sin \frac{\pi}{n}$ diverges.

证明 我们知道,当 $n \to \infty$ 时,$\frac{\ln n}{n} \to 0$. 从而
$$\frac{\ln(n+b)}{n+b} \to 0 \quad (n \to \infty),$$
于是
$$\frac{\ln(n+b)}{n} = \frac{\ln(n+b)}{n+b} \cdot \frac{n+b}{n}$$
$$\to 0 \quad (n \to \infty).$$
因此,对于充分大的 n,有
$$\ln(n+b) < n, \quad 即 \quad \frac{1}{n} < \frac{1}{\ln(n+b)}.$$
由于调和级数 $\sum_{n=1}^{\infty} \frac{1}{n}$ 是发散的,我们得出级数 $\sum_{n=1}^{\infty} \frac{1}{\ln(n+b)}$ 是发散的.

定理 4.4(比较判别法 1) 设 $\sum_{n=1}^{\infty} a_n$ 和 $\sum_{n=1}^{\infty} b_n$ 均为正项级数. 如果
$$\lim_{n \to \infty} \frac{a_n}{b_n} = l,$$
其中 l 是某个正数,那么级数 $\sum_{n=1}^{\infty} a_n$ 和 $\sum_{n=1}^{\infty} b_n$ 同时收敛或发散.

例 4 判断级数 $\sum_{n=1}^{\infty} \sin \frac{\pi}{n}$ 是否收敛或发散.

解 由于当 $x \to 0$ 时,$\frac{\sin x}{x} \to 1$,而当 $n \to \infty$ 时,$\frac{\pi}{n} \to 0$,从而有
$$\frac{\sin \frac{\pi}{n}}{\frac{\pi}{n}} \to 1 \quad (n \to \infty),$$
又知级数 $\sum_{n=1}^{\infty} \frac{\pi}{n}$ 发散(因 $\sum_{n=1}^{\infty} \frac{1}{n}$ 发散),因此级数 $\sum_{n=1}^{\infty} \sin \frac{\pi}{n}$ 是发散的.

Theorem 4.5 (The Comparison Test 2) Let $\sum_{n=1}^{\infty} a_n$ and $\sum_{n=1}^{\infty} b_n$ be positive terms series and
$$\lim_{n \to \infty} \frac{a_n}{b_n} = 0.$$

(1) If $\sum_{n=1}^{\infty} b_n$ converges, and then $\sum_{n=1}^{\infty} a_n$ converges;

(2) If $\sum_{n=1}^{\infty} a_n$ diverges, then $\sum_{n=1}^{\infty} b_n$ diverges;

(3) If $\sum_{n=1}^{\infty} a_n$ converges, then $\sum_{n=1}^{\infty} b_n$ may converge or diverge;

(4) If $\sum_{n=1}^{\infty} b_n$ diverges, then $\sum_{n=1}^{\infty} a_n$ may converge or diverge.

Theorem 4.6 (The Root Test, Cauchy Test) Let $\sum_{n=1}^{\infty} a_n$ be a positive terms series and we have
$$\lim_{n \to \infty} \sqrt[n]{a_n} = \lim a_n^{1/n} = \rho.$$

(1) If $\rho < 1$, then $\sum_{n=1}^{\infty} a_n$ converges;

(2) If $\rho > 1$, then $\sum_{n=1}^{\infty} a_n$ diverges;

(3) If $\rho = 1$, then convergence or divergence of $\sum_{n=1}^{\infty} a_n$ is inconclusive.

Example 5 Determine whether the series $\sum_{n=1}^{\infty} \frac{1}{(\ln n)^n}$ converges or diverges.

Solution For the series $\sum_{n=1}^{\infty} \frac{1}{(\ln n)^n}$, applying the Root Test, we have
$$\lim_{n \to \infty} (a_n)^{1/n} = \lim_{n \to \infty} \frac{1}{\ln n} = 0.$$
So the series converges.

Example 6 Determine whether the series $\sum_{n=1}^{\infty} \frac{2^n}{n^3}$ converges or diverges.

定理 4.5(比较判别法 2) 设 $\sum_{n=1}^{\infty} a_n$ 和 $\sum_{n=1}^{\infty} b_n$ 均为正项级数,并且
$$\lim_{n \to \infty} \frac{a_n}{b_n} = 0.$$

(1) 如果 $\sum_{n=1}^{\infty} b_n$ 收敛,那么 $\sum_{n=1}^{\infty} a_n$ 收敛;

(2) 如果 $\sum_{n=1}^{\infty} a_n$ 发散,那么 $\sum_{n=1}^{\infty} b_n$ 发散;

(3) 如果 $\sum_{n=1}^{\infty} a_n$ 收敛,那么 $\sum_{n=1}^{\infty} b_n$ 可能收敛也可能发散;

(4) 如果 $\sum_{n=1}^{\infty} b_n$ 发散,那么 $\sum_{n=1}^{\infty} a_n$ 可能收敛也可能发散.

定理 4.6(根值判别法,柯西判别法) 设 $\sum_{n=1}^{\infty} a_n$ 为正项级数,并且有
$$\lim_{n \to \infty} \sqrt[n]{a_n} = \lim a_n^{1/n} = \rho.$$

(1) 如果 $\rho < 1$,那么 $\sum_{n=1}^{\infty} a_n$ 收敛;

(2) 如果 $\rho > 1$,那么 $\sum_{n=1}^{\infty} a_n$ 发散;

(3) 如果 $\rho = 1$,那么 $\sum_{n=1}^{\infty} a_n$ 的敛散性不确定.

例 5 判断级数 $\sum_{n=1}^{\infty} \frac{1}{(\ln n)^n}$ 是收敛还是发散.

解 对于级数 $\sum_{n=1}^{\infty} \frac{1}{(\ln n)^n}$,应用根值判别法,我们有
$$\lim_{n \to \infty} (a_n)^{1/n} = \lim_{n \to \infty} \frac{1}{\ln n} = 0.$$
所以该级数收敛.

例 6 判断级数 $\sum_{n=1}^{\infty} \frac{2^n}{n^3}$ 是否收敛或发散.

Solution For the series $\sum_{n=1}^{\infty} \frac{2^n}{n^3}$, applying the Root Test, we have

$$\lim_{n\to\infty} a_n^{1/n} = \lim_{n\to\infty} 2\left(\frac{1}{n}\right)^{3/n}$$
$$= \lim_{n\to\infty} 2\left(\frac{1}{n^{1/n}}\right)^3$$
$$= 2 \times 1^3$$
$$= 2 > 1.$$

Here we use the limit

$$\lim_{n\to\infty} n^{1/n} = \lim_{n\to\infty} \sqrt[n]{n} = 1.$$

So the series diverges.

We continue to consider only positive terms series. The following is the simplest test for convergence and divergence of positive terms series: the Ratio Test.

Theorem 4.7 (The Ratio Test, D'Alembert Test)

Let $\sum_{n=1}^{\infty} a_n$ be a positive terms series and suppose that

$$\lim_{n\to\infty} \frac{a_{n+1}}{a_n} = \lambda.$$

(1) If $\lambda < 1$, then $\sum_{n=1}^{\infty} a_n$ converges;

(2) If $\lambda > 1$, then $\sum_{n=1}^{\infty} a_n$ diverges;

(3) If $\lambda = 1$, then $\sum_{n=1}^{\infty} a_n$ may be convergent or divergent.

Example 7 Prove that the series $\sum_{n=1}^{\infty} \frac{n}{3^n}$ converges.

Proof Let $a_n = \frac{n}{3^n}$, then

$$\frac{a_{n+1}}{a_n} = \frac{n+1}{3^{n+1}} \cdot \frac{3^n}{n}$$
$$= \frac{n+1}{n} \cdot \frac{1}{3}.$$

This ratio approaches $\frac{1}{3}$ as $n \to \infty$, that is

$$\lim_{n\to\infty} \frac{a_{n+1}}{a_n} = \frac{1}{3}.$$

解 对于级数 $\sum_{n=1}^{\infty} \frac{2^n}{n^3}$，应用根值判别法，我们有

$$\lim_{n\to\infty} a_n^{1/n} = \lim_{n\to\infty} 2\left(\frac{1}{n}\right)^{3/n}$$
$$= \lim_{n\to\infty} 2\left(\frac{1}{n^{1/n}}\right)^3$$
$$= 2 \times 1^3$$
$$= 2 > 1,$$

这里用到了极限

$$\lim_{n\to\infty} n^{1/n} = \lim_{n\to\infty} \sqrt[n]{n} = 1.$$

因此该级数发散．

我们继续考虑正项级数．下面介绍关于正项级数敛散性的最简单的判别法——比值判别法．

定理 4.7（比值判别法，达朗贝尔判别法）

设 $\sum_{n=1}^{\infty} a_n$ 为正项级数，并且有

$$\lim_{n\to\infty} \frac{a_{n+1}}{a_n} = \lambda.$$

（1）如果 $\lambda < 1$，那么 $\sum_{n=1}^{\infty} a_n$ 收敛；

（2）如果 $\lambda > 1$，那么 $\sum_{n=1}^{\infty} a_n$ 发散；

（3）如果 $\lambda = 1$，那么 $\sum_{n=1}^{\infty} a_n$ 可能收敛，也可能发散．

例 7 证明：级数 $\sum_{n=1}^{\infty} \frac{n}{3^n}$ 收敛．

证明 这里 $a_n = \frac{n}{3^n}$，于是

$$\frac{a_{n+1}}{a_n} = \frac{n+1}{3^{n+1}} \cdot \frac{3^n}{n}$$
$$= \frac{n+1}{n} \cdot \frac{1}{3}.$$

当 $n \to \infty$ 时，这个比值趋近于 $\frac{1}{3}$，即

$$\lim_{n\to\infty} \frac{a_{n+1}}{a_n} = \frac{1}{3},$$

Hence the Ratio Test is applicable, and the series converges.

Example 8 Prove that the series $\sum_{n=1}^{\infty} \dfrac{n^n}{n!}$ diverges.

Proof We have
$$\frac{a_{n+1}}{a_n} = \frac{(n+1)^{n+1}}{(n+1)!} \cdot \frac{n!}{n^n}$$
$$= \left(\frac{n+1}{n}\right)^n$$
$$= \left(1 + \frac{1}{n}\right)^n,$$
so
$$\lim_{n \to \infty} \frac{a_{n+1}}{a_n} = \lim_{n \to \infty} \left(1 + \frac{1}{n}\right)^n = e.$$
Since $e > 1$, the series diverges.

4.3 Alternating Series, Absolute Convergence and Conditional Convergence

In this section we consider series which have both positive and negative terms.

1. Alternating Series and Its Tests for Convergence

Definition 4.3 The series, such as,
$$u_1 - u_2 + u_3 - u_4 + \cdots$$
or
$$-u_1 + u_2 - u_3 + u_4 - \cdots$$
is called the **alternating series**, where $u_n > 0$ for any n.

For example, the following two series are both alternating series:
$$1 - \frac{1}{2} + \frac{1}{3} - \frac{1}{4} + \frac{1}{5} - \frac{1}{6} + \cdots$$
$$= \sum_{n=1}^{\infty} \frac{(-1)^{n-1}}{n},$$
$$-\frac{1}{2} + \frac{2}{3} - \frac{3}{4} + \frac{4}{5} - \frac{5}{6} + \cdots$$
$$= \sum_{n=1}^{\infty} \frac{(-1)^n}{n+1}.$$

因此由比值判别法知,该级数收敛.

例 8 证明:级数 $\sum_{n=1}^{\infty} \dfrac{n^n}{n!}$ 发散.

证明 我们有
$$\frac{a_{n+1}}{a_n} = \frac{(n+1)^{n+1}}{(n+1)!} \cdot \frac{n!}{n^n}$$
$$= \left(\frac{n+1}{n}\right)^n$$
$$= \left(1 + \frac{1}{n}\right)^n,$$
因此
$$\lim_{n \to \infty} \frac{a_{n+1}}{a_n} = \lim_{n \to \infty} \left(1 + \frac{1}{n}\right)^n = e.$$
因为 $e > 1$,所以该级数发散.

4.3 交错级数,绝对收敛和条件收敛

这一节我们考虑既有正项也有负项的级数.

1. 交错级数及其收敛判别法

定义 4.3 形如
$$u_1 - u_2 + u_3 - u_4 + \cdots$$
或
$$-u_1 + u_2 - u_3 + u_4 - \cdots$$
的级数称为**交错级数**,其中对于任意 n,有 $u_n > 0$.

例如,下面两个级数均为交错级数:
$$1 - \frac{1}{2} + \frac{1}{3} - \frac{1}{4} + \frac{1}{5} - \frac{1}{6} + \cdots$$
$$= \sum_{n=1}^{\infty} \frac{(-1)^{n-1}}{n},$$
$$-\frac{1}{2} + \frac{2}{3} - \frac{3}{4} + \frac{4}{5} - \frac{5}{6} + \cdots$$
$$= \sum_{n=1}^{\infty} \frac{(-1)^n}{n+1}.$$

4.3 Alternating Series, Absolute Convergence and Conditional Convergence

We see from these examples that the nth term of an alternating series is
$$a_n = (-1)^{n-1} u_n \quad \text{or} \quad a_n = (-1)^n u_n,$$
where u_n is positive for any n (in fact $u_n = |a_n|$).

The following test says that if the absolute value of the terms of an alternating series decrease toward 0, then the series converges.

Theorem 4.8 (Leibniz Test) If the alternating series $\sum_{n=1}^{\infty} (-1)^n u_n \ (u_n > 0, n \geq 1)$ satisfies:

(1) $u_n \geq u_{n+1}, n \geq 1$;

(2) $\lim\limits_{n \to \infty} u_n = 0$,

then the series converges. Moreover, its sum $S \leq u_1$, and the error $|r_n|$ make by using the sum S_n of the first n terms to approximate the sum S of the series is not more than u_{n+1}, that is
$$|r_n| = |S - S_n| \leq u_{n+1}.$$

Example 1 Prove that the following **alternating harmonic series** is convergent:
$$1 - \frac{1}{2} + \frac{1}{3} - \frac{1}{4} + \cdots = \sum_{n=1}^{\infty} \frac{(-1)^{n-1}}{n}.$$

Proof The alternating harmonic series satisfies:

(1) $u_{n+1} = \dfrac{1}{n+1} < u_n = \dfrac{1}{n} \ (n \geq 1)$;

(2) $\lim\limits_{n \to \infty} u_n = \lim\limits_{n \to \infty} \dfrac{1}{n} = 0$.

So the series is convergent by Leibniz Test.

Example 2 Test the series $\sum_{n=1}^{\infty} \dfrac{(-1)^n 3n}{4n-1}$ for convergence or divergence.

Solution The given series is alternating but
$$\lim_{n \to \infty} u_n = \lim_{n \to \infty} \frac{3n}{4n-1} = \lim_{n \to \infty} \frac{3}{4 - \frac{1}{n}}$$
$$= \frac{3}{4} \neq 0.$$

So condition (2) of Leibniz Test is not satisfied.

Instead, we look at the limit of the nth term of the series:

从这两个例子我们看出, 交错级数的第 n 项的形式为
$$a_n = (-1)^{n-1} u_n \quad \text{或} \quad a_n = (-1)^n u_n,$$
其中对任意 n, u_n 是一个正数 (事实上, $u_n = |a_n|$).

下面的判别法说明, 如果一个交错级数的项的绝对值递减趋于 0, 那么这个级数收敛.

定理 4.8 (莱布尼茨判别法) 如果交错级数 $\sum_{n=1}^{\infty} (-1)^n u_n \ (u_n > 0, n \geq 1)$ 满足:

(1) $u_n \geq u_{n+1}, n \geq 1$;

(2) $\lim\limits_{n \to \infty} u_n = 0$,

那么这个级数收敛. 此外, 其和 $S \leq u_1$, 并且用前 n 项部分和 S_n 去近似整个级数的和 S 的误差 $|r_n|$ 比 u_{n+1} 小, 即
$$|r_n| = |S - S_n| \leq u_{n+1}.$$

例 1 证明下面的**交错调和级数**收敛:
$$1 - \frac{1}{2} + \frac{1}{3} - \frac{1}{4} + \cdots = \sum_{n=1}^{\infty} \frac{(-1)^{n-1}}{n}.$$

证明 该交错调和级数满足:

(1) $u_{n+1} = \dfrac{1}{n+1} < u_n = \dfrac{1}{n} \ (n \geq 1)$;

(2) $\lim\limits_{n \to \infty} u_n = \lim\limits_{n \to \infty} \dfrac{1}{n} = 0$.

因此, 根据莱布尼茨判别法, 该级数收敛.

例 2 判断级数 $\sum_{n=1}^{\infty} \dfrac{(-1)^n 3n}{4n-1}$ 是收敛还是发散.

解 给出的级数是交错级数, 但是
$$\lim_{n \to \infty} u_n = \lim_{n \to \infty} \frac{3n}{4n-1} = \lim_{n \to \infty} \frac{3}{4 - \frac{1}{n}}$$
$$= \frac{3}{4} \neq 0,$$

因此莱布尼茨判别法的条件 (2) 不满足.

我们转而看该级数第 n 项的极限:

$$\lim_{n\to\infty}a_n = \lim_{n\to\infty}\frac{(-1)^n 3n}{4n-1}.$$

When n is large enough and even, a_n is close to $\frac{3}{4}$; when n is large enough and odd, a_n is close to $-\frac{3}{4}$. Since the limit $\lim_{n\to\infty}a_n$ does not exist. The series is divergent.

2. Absolute and Conditional Convergence

Now, we consider general series which have both positive and negative terms but are not alternating series. For general series $\sum_{n=1}^{\infty}a_n$, in order to make use of convergence or divergence test of positive terms series. The term of the series are usually added in absolute terms, considering the series $\sum_{n=1}^{\infty}|a_n|$. To the end, the concepts of absolute convergence and conditional convergence are introduced.

Definition 4.4 If the series $\sum_{n=1}^{\infty}|a_n|$ converges, then the series $\sum_{n=1}^{\infty}a_n$ is called **absolutely convergent**. The series $\sum_{n=1}^{\infty}a_n$ is called **conditionally convergent**, if the series $\sum_{n=1}^{\infty}a_n$ converges, but $\sum_{n=1}^{\infty}|a_n|$ diverges.

Theorem 4.9 If the series $\sum_{n=1}^{\infty}|a_n|$ converges, then the series $\sum_{n=1}^{\infty}a_n$ converges.

This theorem merely states that absolutely convergent series are convergent. As we will show presently, the converse is false. There are convergent series that are not absolutely convergent, such series are called conditionally convergent as the series in Example 5.

Example 3 Prove that the following series is absolutely convergent:
$$1 - \frac{1}{2^2} + \frac{1}{3^2} - \frac{1}{4^2} + \frac{1}{5^2} + \cdots.$$

4.3 Alternating Series, Absolute Convergence and Conditional Convergence
4.3 交错级数,绝对收敛和条件收敛

Proof If we replace each term by its absolute value, we obtain the series

$$1+\frac{1}{2^2}+\frac{1}{3^2}+\frac{1}{4^2}+\cdots.$$

This is a p-series with $p=2$. It is therefore convergent. This means that the initial series is absolutely convergent.

Example 4 Prove that the following series is absolutely convergent:

$$1-\frac{1}{2}-\frac{1}{2^2}+\frac{1}{2^3}-\frac{1}{2^4}-\frac{1}{2^5}$$
$$+\frac{1}{2^6}-\frac{1}{2^7}-\frac{1}{2^8}+\cdots.$$

Proof If we replace each term by its absolute value, we obtain the series:

$$1+\frac{1}{2}+\frac{1}{2^2}+\frac{1}{2^3}+\frac{1}{2^4}+\frac{1}{2^5}$$
$$+\frac{1}{2^6}+\frac{1}{2^7}+\frac{1}{2^8}+\cdots$$
$$=\sum_{n=0}^{\infty}\left(\frac{1}{2}\right)^n.$$

This is a convergent geometric series. The initial series is therefore absolutely convergent.

Example 5 Prove that the following series is conditionally convergent:

$$-1+\frac{1}{2}-\frac{1}{3}+\frac{1}{4}-\frac{1}{5}+\frac{1}{6}+\cdots$$
$$=\sum_{n=1}^{\infty}\frac{(-1)^n}{n}.$$

Proof The given series is the alternating series which satisfies:

(1) $u_{n+1}=\dfrac{1}{n+1}<u_n=\dfrac{1}{n}$ $(n\geqslant 1)$;

(2) $\lim\limits_{n\to\infty}u_n=\lim\limits_{n\to\infty}\dfrac{1}{n}=0.$

So this series is convergent by the Leibniz Test, but it is not absolutely convergent. In fact, if we replace each term by its absolute value, we obtain the divergent harmonic series:

$$1+\frac{1}{2}+\frac{1}{3}+\frac{1}{4}+\cdots=\sum_{n=1}^{\infty}\frac{1}{n}.$$

So the given series is conditionally convergent.

证明 如果用各项的绝对值代替每一项,我们获得级数

$$1+\frac{1}{2^2}+\frac{1}{3^2}+\frac{1}{4^2}+\cdots.$$

这是一个 p 级数,且 $p=2$,因此它收敛.这意味着原级数绝对收敛.

例 4 证明下面的级数绝对收敛:

$$1-\frac{1}{2}-\frac{1}{2^2}+\frac{1}{2^3}-\frac{1}{2^4}-\frac{1}{2^5}$$
$$+\frac{1}{2^6}-\frac{1}{2^7}-\frac{1}{2^8}+\cdots.$$

证明 如果用各项的绝对值代替每一项,我们获得级数

$$1+\frac{1}{2}+\frac{1}{2^2}+\frac{1}{2^3}+\frac{1}{2^4}+\frac{1}{2^5}$$
$$+\frac{1}{2^6}+\frac{1}{2^7}+\frac{1}{2^8}+\cdots$$
$$=\sum_{n=0}^{\infty}\left(\frac{1}{2}\right)^n.$$

这是一个收敛的几何级数.因此原级数是绝对收敛的.

例 5 证明下面的级数条件收敛:

$$-1+\frac{1}{2}-\frac{1}{3}+\frac{1}{4}-\frac{1}{5}+\frac{1}{6}+\cdots$$
$$=\sum_{n=1}^{\infty}\frac{(-1)^n}{n}.$$

证明 给定级数是交错级数,它满足条件:

(1) $u_{n+1}=\dfrac{1}{n+1}<u_n=\dfrac{1}{n}$ $(n\geqslant 1)$;

(2) $\lim\limits_{n\to\infty}u_n=\lim\limits_{n\to\infty}\dfrac{1}{n}=0.$

因此,由莱布尼兹判别法知这个级数收敛.但它不是绝对收敛的.事实上,如果用各项的绝对值代替每一项,我们获得发散的调和级数:

$$1+\frac{1}{2}+\frac{1}{3}+\frac{1}{4}+\cdots=\sum_{n=1}^{\infty}\frac{1}{n}.$$

因此,给定的级数是条件收敛的.

4.4 Power Series
4.4 幂级数

Definition 4.5 The formula
$$u_1(x)+u_2(x)+\cdots+u_n(x)+\cdots \quad (4.5)$$
is called the **infinite series of functions** on the interval I, referred to as **series of function** or **series**, where
$$u_1(x), u_2(x), \cdots, u_n(x), \cdots$$
are all functions defined on the interval I. For the series of function (4.5), the point x_0 is called the **point of convergece** if the series
$$u_1(x_0)+u_2(x_0)+\cdots+u_n(x_0)+\cdots$$
is convergent. We call the set of convergence points of the series of functions (4.5) the **domain of convergence** of the series (4.5).

Let D be the domain of convergence of the series of function (4.5), then for any $x \in D$, there is a unique sum $S = \sum_{n=1}^{\infty} u_n(x)$ corresponding to it. In this way, a new function is defined on D, which we call the **sum function** of series (4.5), denoted as $S(x)$, that is
$$S(x) = \sum_{n=1}^{\infty} u_n(x)$$
$$= u_1(x) + u_2(x) + \cdots + u_n(x) + \cdots,$$
where D is the domain of $S(x)$.

Remember that the partial sum of the first n terms of the series of function is $S_n(x)$, that is
$$S_n(x) = \sum_{k=1}^{n} u_k(x)$$
$$= u_1(x) + u_2(x) + \cdots + u_n(x),$$
Obviously in the convergence domain D,
$$\lim_{n \to \infty} S_n(x) = S(x).$$
Similar to the series of constant, we say that
$$r_n(x) = S(x) - S_n(x)$$
is the **remainder** of series of function (4.5), and obviously

定义 4.5 形如
$$u_1(x)+u_2(x)+\cdots+u_n(x)+\cdots \quad (4.5)$$
的级数称为区间 I 上的**函数项无穷级数**，简称**函数项级数**或**级数**，这里
$$u_1(x), u_2(x), \cdots, u_n(x), \cdots$$
为定义在区间 I 上的函数. 对于函数项级数 (4.5)，点 x_0 称为**收敛点**，如果级数
$$u_1(x_0)+u_2(x_0)+\cdots+u_n(x_0)+\cdots$$
收敛. 我们称函数项级数 (4.5) 所有收敛点组成的集合为级数 (4.5) 的**收敛域**.

设 D 为函数项级数 (4.5) 的收敛域，则对于任意的 $x \in D$，都有唯一的和 $S = \sum_{n=1}^{\infty} u_n(x)$ 与之对应. 这样就在 D 上定义了一个新函数，我们称之为函数项级数 (4.5) 的**和函数**，记为 $S(x)$，即
$$S(x) = \sum_{n=1}^{\infty} u_n(x)$$
$$= u_1(x) + u_2(x) + \cdots + u_n(x) + \cdots,$$
其中 D 就是 $S(x)$ 的定义域.

记函数项级数 (4.5) 的前 n 项部分和为 $S_n(x)$，即
$$S_n(x) = \sum_{k=1}^{n} u_k(x)$$
$$= u_1(x) + u_2(x) + \cdots + u_n(x),$$
显然在收敛域 D 上有
$$\lim_{n \to \infty} S_n(x) = S(x).$$
类似于常数项级数的情形，我们称为
$$r_n(x) = S(x) - S_n(x)$$
为函数项级数 (4.5) 的**余项**，且显然有

$$\lim_{n\to\infty} r_n(x) = 0, \quad x \in D.$$

A kind of the special series of function is power series. Generally, a series of the form

$$\sum_{n=0}^{\infty} a_n(x-a)^n = a_0 + a_1(x-a) + a_2(x-a)^2 + \cdots \quad (4.6)$$

is called a **power series**. When $x=a$, all of the terms are 0 for $n \geq 1$ and so the power series (4.6) always converges.

We can obtain a special power series when $a=0$. Since a simple translation converts $\sum_{n=0}^{\infty} a_n(x-a)^n$ into

$$\sum_{n=0}^{\infty} a_n x^n = a_0 + a_1 x + a_2 x^2 + \cdots,$$

we can focus our attention on the power series $\sum_{n=0}^{\infty} a_n x^n$.

For the power series $\sum_{n=1}^{\infty} a_n x^n$, the following result is fundamental.

Theorem 4.10 (Abel Theorem)

(1) If the power series $\sum_{n=0}^{\infty} a_n x^n$ converges at $x_1 \neq 0$, it is absolutely convergent, for any x that $|x| < |x_1|$;

(2) If the power series $\sum_{n=0}^{\infty} a_n x^n$ diverges at $x_1 \neq 0$, it diverges for any x that $|x| > |x_1|$.

Corollary If the domain of convergence for a power series $\sum_{n=0}^{\infty} a_n x^n$ is neither $\{x=0\}$ nor the real number axis, then there exists a positive number R, such that

(1) if $|x| < R$, $\sum_{n=0}^{\infty} a_n x^n$ is absolutely convergent;

(2) if $|x| > R$, $\sum_{n=0}^{\infty} a_n x^n$ is divergent;

(3) if $x=R$ and $x=-R$, $\sum_{n=0}^{\infty} a_n x^n$ may be convergent or divergent.

$$\lim_{n\to\infty} r_n(x) = 0, \quad x \in D.$$

一类特殊的函数项级数就是幂级数. 一般地, 我们称形如

$$\sum_{n=0}^{\infty} a_n(x-a)^n = a_0 + a_1(x-a) + a_2(x-a)^2 + \cdots \quad (4.6)$$

的级数为**幂级数**. 当 $x=a$ 时, 对于 $n \geq 1$ 的所有项都为零, 此时幂级数(4.6)总是收敛的.

当 $a=0$ 时, 我们得到特殊的幂级数

$$\sum_{n=0}^{\infty} a_n x^n = a_0 + a_1 x + a_2 x^2 + \cdots.$$

因为通过一个简单的变换, 就能将幂级数 $\sum_{n=0}^{\infty} a_n(x-a)^n$ 变成级数 $\sum_{n=0}^{\infty} a_n x^n$, 从而我们可以把注意力集中在研究幂级数 $\sum_{n=0}^{\infty} a_n x^n$ 上.

对于幂级数 $\sum_{n=0}^{\infty} a_n x^n$, 下面的结果是基本的.

定理 4.10(阿贝尔定理)

(1) 若幂级数 $\sum_{n=0}^{\infty} a_n x^n$ 在 $x_1 \neq 0$ 处收敛, 则对任何满足 $|x| < |x_1|$ 的 x, 该幂级数绝对收敛;

(2) 若幂级数 $\sum_{n=0}^{\infty} a_n x^n$ 在 $x_1 \neq 0$ 处发散, 则对任何满足 $|x| > |x_1|$ 的 x, 该幂级数发散.

推论 如果幂级数 $\sum_{n=0}^{\infty} a_n x^n$ 的收敛域不是 $\{x=0\}$, 也不是整个实数轴, 那么存在一个正数 R, 满足:

(1) 若 $|x| < R$, 则 $\sum_{n=0}^{\infty} a_n x^n$ 绝对收敛;

(2) 若 $|x| > R$, 则 $\sum_{n=0}^{\infty} a_n x^n$ 是发散;

(3) 若 $x=R$ 和 $x=-R$, 则 $\sum_{n=0}^{\infty} a_n x^n$ 或者收敛或者发散.

The positive number R satisfied above conditions (1) and (2) is called the **radius of convergence** of the power series $\sum_{n=0}^{\infty} a_n x^n$.

In general, for a given power series $\sum_{n=0}^{\infty} a_n (x-a)^n$, if there exists a positive number R such that the series converges if $|x-a|<R$ and diverges if $|x-a|>R$. The positive number R is called the **radius of convergence** of the power series $\sum_{n=0}^{\infty} a_n (x-a)^n$.

From the theorem given above we can see that there are three possibility cases for the power series $\sum_{n=0}^{\infty} a_n x^n$.

Case 1 The power series $\sum_{n=0}^{\infty} a_n x^n$ converges only at $x=0$. This is what happens with the power series $\sum_{n=1}^{\infty} n^n x^n$. Now we say that the radius of convergence is 0.

Case 2 The power series $\sum_{n=0}^{\infty} a_n x^n$ is absolutely convergent everywhere. This is what happens with the power series $\sum_{n=1}^{\infty} \frac{x^n}{n!}$. Now we say that the radius of convergence is ∞.

Case 3 There exists a positive number R such that the power series $\sum_{n=0}^{\infty} a_n x^n$ is absolutely convergent for $|x|<R$ and diverges for $|x|>R$. In this case, the radius of convergence is R. This is what happens with the geometric series $\sum_{n=0}^{\infty} x^n$. It is absolutely convergent for $|x|<1$, and divergent for $|x|>1$. Thus the radius of convergence of this power series is 1.

If the radius of convergence of the power series $\sum_{n=0}^{\infty} a_n x^n$ is R, then we say that $(-R, R)$ is **convergence interval** of this power series. In particular, the convergence interval shrinks to $\{0\}$ when $R=0$.

Generally speaking, the convergence of the end points

满足上述条件(1)和(2)的正数 R 称为幂级数 $\sum_{n=0}^{\infty} a_n x^n$ 的**收敛半径**.

一般地,对给定的幂级数 $\sum_{n=0}^{\infty} a_n (x-a)^n$,如果存在一个正数 R,当 $|x-a|<R$ 时,幂级数收敛;当 $|x-a|>R$ 时,幂级数发散,则称正数 R 为幂级数 $\sum_{n=0}^{\infty} a_n (x-a)^n$ 的**收敛半径**.

从上面给出的定理我们可以看出,对于幂级数 $\sum_{n=0}^{\infty} a_n x^n$,恰好有三种可能情形:

情形 1 幂级数 $\sum_{n=0}^{\infty} a_n x^n$ 只在点 $x=0$ 收敛,比如幂级数 $\sum_{n=0}^{\infty} n^n x^n$. 这时,我们说收敛半径是 0.

情形 2 幂级数 $\sum_{n=0}^{\infty} a_n x^n$ 处处绝对收敛,比如幂级数 $\sum_{n=1}^{\infty} \frac{x^n}{n!}$. 这时,我们说收敛半径是 ∞.

情形 3 存在一个正数 R,使得幂级数 $\sum_{n=0}^{\infty} a_n x^n$ 对于 $|x|<R$ 绝对收敛,而对于 $|x|>R$ 发散. 这时,收敛半径为 R.

例如,几何级数 $\sum_{n=0}^{\infty} x^n$ 对于 $|x|<1$ 绝对收敛,对于 $|x|>1$ 发散,所以该幂级数的收敛半径为 1.

若幂级数 $\sum_{n=0}^{\infty} a_n x^n$ 的收敛半径为 R,则称 $(-R, R)$ 为该幂级数的**收敛区间**. 特别地,当 $R=0$ 时,收敛区间缩为 $\{0\}$.

一般来说,幂级数 $\sum_{n=0}^{n} a_n x^n$ 在收敛区间

$x=-R$ and $x=R$ of power series $\sum_{n=0}^{n} a_n x^n$ in interval of convergence can not be decided ture, it may or may not converge. For example, the power series $\sum_{n=1}^{\infty} x^n$, $\sum_{n=1}^{\infty} \frac{x^n}{n}$, $\sum_{n=1}^{\infty} \frac{(-1)^n}{n} x^n$, $\sum_{n=1}^{\infty} \frac{1}{n^2} x^n$ all have the radius of convergence 1, but while the first power series converges only on $(-1,1)$, the second series converges on $[-1,1)$, the third on $(-1,1]$, and the forth on $[-1,1]$.

For a power series $\sum_{n=0}^{\infty} a_n x^n$ with the radius of convergence R, the intervals of convergence have four cases. They may be $[-R,R], [-R,R), (-R,R]$ or $(-R,R)$. For a power series with the radius of convergence 0, its interval of convergence reduces to a point $\{0\}$.

In the following we give these methods of finding the radius of convergence.

Theorem 4.11 For the power series $\sum_{n=0}^{\infty} a_n x^n$, if
$$\lim_{n\to\infty}\left|\frac{a_{n+1}}{a_n}\right|=\rho,$$
the radius of convergence of the power series is
$$R=\begin{cases} \frac{1}{\rho}, & \rho \neq 0, \\ +\infty, & \rho = 0, \\ 0, & \rho = +\infty. \end{cases}$$

Theorem 4.12 For the power series $\sum_{n=0}^{\infty} a_n x^n$, if
$$\lim_{n\to\infty} \sqrt[n]{a_n} = \lim_{n\to\infty} |a_n|^{\frac{1}{n}} = \rho,$$
the radius of convergence of the power series is
$$R=\begin{cases} \frac{1}{\rho}, & \rho \neq 0 \\ +\infty, & \rho = 0 \\ 0, & \rho = +\infty. \end{cases}$$

Example 1 Verify that the power series $\sum_{n=1}^{\infty} \frac{(-1)^n}{n} x^n$ has the interval of convergence $(-1,1]$.

Solution First we use the theorem 4.11 to solve the radius of convergence. Let

$$a_n = \frac{(-1)^n}{n}, \quad a_{n+1} = \frac{(-1)^{n+1}}{n+1},$$

thus we have

$$\rho = \lim_{n\to\infty}\left|\frac{a_{n+1}}{a_n}\right| = \lim_{n\to\infty}\frac{n}{n+1}$$

$$= \lim_{n\to\infty}\frac{\frac{n}{n}}{1+\frac{1}{n}} = 1 \neq 0.$$

By Theorem 4.11, the radius of convergence is

$$R = \frac{1}{\rho} = 1.$$

It follows that the given series is absolutely convergent for $|x|<1$ and divergent for $|x|>1$.

Now we test either convergence or divergence at endpoints $x=-1$ and $x=1$. At $x=-1$, $\sum_{n=1}^{\infty}\frac{(-1)^n}{n}x^n$ becomes

$$\sum_{n=1}^{\infty}\frac{(-1)^n}{n}(-1)^n = \sum_{n=1}^{\infty}\frac{1}{n}.$$

This is a harmonic series which diverges. At $x = 1$, $\sum_{n=n}^{\infty}\frac{(-1)^n}{n}x^n$ becomes

$$\sum_{n=1}^{\infty}\frac{(-1)^n}{n}.$$

This is a convergent alternating series. Therefore the interval of convergence of power series is $(-1,1]$.

Example 2 Find the convergence domain of power series $\sum_{n=1}^{\infty}\frac{(x-1)^n}{2^n n}$.

Solution This power series is not the form as $\sum_{n=0}^{\infty}a_n x^n$, but we let $t = x-1$, the power series could change to $\sum_{n=1}^{\infty}\frac{t^n}{2^n n}$. For this new power series, because

$$\rho = \lim_{n\to\infty}\left|\frac{a_{n+1}}{a_n}\right| = \lim_{n\to\infty}\left|\frac{\frac{1}{2^{n+1}(n+1)}}{\frac{1}{2^n n}}\right|$$

$$= \lim_{n\to\infty}\frac{n}{n+1}\cdot\frac{2^n}{2^{n+1}}$$

$$a_n = \frac{(-1)^n}{n}, \quad a_{n+1} = \frac{(-1)^{n+1}}{n+1},$$

所以

$$\rho = \lim_{n\to\infty}\left|\frac{a_{n+1}}{a_n}\right| = \lim_{n\to\infty}\frac{n}{n+1}$$

$$= \lim_{n\to\infty}\frac{\frac{n}{n}}{1+\frac{1}{n}} = 1 \neq 0.$$

由定理4.11,收敛半径为

$$R = \frac{1}{\rho} = 1.$$

因此,给出的级数对于$|x|<1$绝对收敛,对于$|x|>1$发散.

现在我们检验在端点$x=-1$和$x=1$处的敛散性. 在$x=-1$时,$\sum_{n=1}^{\infty}\frac{(-1)^n}{n}x^n$可化为

$$\sum_{n=1}^{\infty}\frac{(-1)^n}{n}(-1)^n = \sum_{n=1}^{\infty}\frac{1}{n}.$$

这是调和级数,它是发散的. 在$x=1$时,$\sum_{n=n}^{\infty}\frac{(-1)^n}{n}x^n$可化为

$$\sum_{n=1}^{\infty}\frac{(-1)^n}{n}.$$

这是一个收敛的交错级数.因此,原幂级数的收敛域是$(-1,1]$.

例2 求幂级数$\sum_{n=1}^{\infty}\frac{(x-1)^n}{2^n n}$的收敛域.

解 该幂级数不是$\sum_{n=0}^{\infty}a_n x^n$的形式,但是若令$t=x-1$,则它可化为$\sum_{n=1}^{\infty}\frac{t^n}{2^n n}$. 对于这一新幂级数,因为

$$\rho = \lim_{n\to\infty}\left|\frac{a_{n+1}}{a_n}\right| = \lim_{n\to\infty}\left|\frac{\frac{1}{2^{n+1}(n+1)}}{\frac{1}{2^n n}}\right|$$

$$= \lim_{n\to\infty}\frac{n}{n+1}\cdot\frac{2^n}{2^{n+1}}$$

$$= \lim_{n \to \infty} \frac{n}{2(n+1)} = \frac{1}{2},$$

so the radius of convergence of the new power series is $R = 2$. That is the new power series is absolutely convergent when $|t| < 2$, and divergent when $|t| > 2$. Then the original power series is absolutely convergent when $|x-1| < 2$ ($-1 < x < 3$), and is divergent when $|x-1| > 2$ ($x < -1$ or $x > 3$).

Consider the convergence and divergence of the origin power series at points $x = -1$ and $x = 3$. When $x = -1$, the origin power series becomes

$$\sum_{n=1}^{\infty} \frac{(-1)^n}{n},$$

which is a convergent alternating series (see Example 5 in §4.3). When $x = 3$, the origin power series becomes

$$\sum_{n=1}^{\infty} \frac{1}{n},$$

which is a harmonic series, and it must be divergent.

Thus, the domain of convergence is $[-1, 3)$.

Example 3 Find the domain of convergence for

$$\sum_{n=1}^{\infty} \frac{(x-2)^{2n}}{4^n n}.$$

Solution Let $u_n(x) = \frac{(x-2)^{2n}}{4^n n}$, then

$$\lim_{n \to \infty} \frac{|u_{n+1}(x)|}{|u_n(x)|}$$

$$= \lim_{n \to \infty} \left[\frac{(x-2)^{2(n+1)}}{(n+1) 4^{n+1}} \cdot \frac{4^n n}{(x-2)^{2n}} \right]$$

$$= \frac{(x-2)^2}{4}.$$

If and only if $0 < x < 4$, $\frac{(x-2)^2}{4} < 1$. By the Ratio Test, the given series is absolutely convergent, therefore it is convergent, and it is divergent when $x < 0$ or $x > 4$.

The Ratio Test gives no information when $\frac{(x-2)^2}{4} = 1$, that is when $x = 0$ or $x = 4$, so we must consider $x = 0$ and $x = 4$ separately. If we put $x = 0$ into the power series, this power series will become the harmonic series

$$\sum_{n=1}^{\infty} \frac{1}{n},$$

which is divergent. If $x=4$, the power series is also

$$\sum_{n=1}^{\infty} \frac{1}{n}.$$

Thus, the given power series converges for $0 < x < 4$, namely the domain of convergence for the given power series is the interval $(0, 4)$.

它是发散的. 若 $x=4$, 该幂级数也是

$$\sum_{n=1}^{\infty} \frac{1}{n},$$

因此, 给定的幂级数对于 $0 < x < 4$ 是收敛的, 即它的收敛域是区间 $(0,4)$.

4.5 Operations and Properties of Power Series
4.5 幂级数的运算与性质

1. Operations of Power Series

Let the interval of convergence of the power series

$$\sum_{n=0}^{\infty} a_n x^n \quad \text{and} \quad \sum_{n=0}^{\infty} b_n x^n$$

is $(-R_a, R_a)$ and $(-R_b, R_b)$, respectively, and write $R = \min\{R_a, R_b\}$. According to the properties of the convergent series, we could do the following arithmetic operation:

Addition:

$$\sum_{n=0}^{\infty} a_n x^n + \sum_{n=0}^{\infty} b_n x^n = \sum_{n=0}^{\infty} (a_n + b_n) x^n,$$
$$x \in (-R, R);$$

Subtraction:

$$\sum_{n=0}^{\infty} a_n x^n - \sum_{n=0}^{\infty} b_n x^n = \sum_{n=0}^{\infty} (a_n - b_n) x^n,$$
$$x \in (-R, R);$$

Multiplication:

$$\sum_{n=0}^{\infty} a_n x^n \cdot \sum_{n=0}^{\infty} b_n x^n = \sum_{n=0}^{\infty} c_n x^n,$$
$$x \in (-R, R),$$

where

$$c_n = a_0 b_n + a_1 b_{n-1} + \cdots + a_{n-1} b_1 + a_n b_0.$$

2. Properties of Power Series

According to the calculus of function of one variable,

1. 幂级数的运算

设幂级数

$$\sum_{n=0}^{\infty} a_n x^n \quad 和 \quad \sum_{n=0}^{\infty} b_n x^n$$

的收敛区间分别为 $(-R_a, R_a)$ 和 $(-R_b, R_b)$, 并记 $R = \min\{R_a, R_b\}$. 根据收敛级数的性质, 可以进行如下四则运算:

加法运算:

$$\sum_{n=0}^{\infty} a_n x^n + \sum_{n=0}^{\infty} b_n x^n = \sum_{n=0}^{\infty} (a_n + b_n) x^n,$$
$$x \in (-R, R);$$

减法运算:

$$\sum_{n=0}^{\infty} a_n x^n - \sum_{n=0}^{\infty} b_n x^n = \sum_{n=0}^{\infty} (a_n - b_n) x^n,$$
$$x \in (-R, R);$$

乘法运算:

$$\sum_{n=0}^{\infty} a_n x^n \cdot \sum_{n=0}^{\infty} b_n x^n = \sum_{n=0}^{\infty} c_n x^n,$$
$$x \in (-R, R),$$

其中

$$c_n = a_0 b_n + a_1 b_{n-1} + \cdots + a_{n-1} b_1 + a_n b_0.$$

2. 幂级数的性质

从一元函数微积分可知, 有限项在区间

the sum of finite functions which are continuous, derivative and integral on interval I is also continuous, derivative and integral. These conclusions can be generalized to the sum of infinite power functions.

Property 1 (Continuity of the Sum Function) Suppose that the convergence domain of the power series $\sum_{n=0}^{\infty} a_n x^n$ is $D \neq \{0\}$, then the sum function $S(x)$ of the power series is continuous in D.

Property 2 (Integrate Term by Term) Suppose that the convergence interval of the power series $\sum_{n=0}^{\infty} a_n x^n$ is $(-R, R)$, then the sum function $S(x)$ of the power series is integrable in $(-R, R)$, and for any $x \in (-R, R)$, we have

$$\int_0^x S(x) dx = \int_0^x \left(\sum_{n=0}^{\infty} a_n x^n \right) dx$$
$$= \sum_{n=0}^{\infty} \int_0^x a_n x^n dx$$
$$= \sum_{n=0}^{\infty} \frac{a_n}{n+1} x^{n+1}$$
$$= a_0 x + \frac{1}{2} a_1 x^2 + \frac{1}{3} a_2 x^3 + \cdots,$$

where the convergence radius of the power series $\sum_{n=0}^{\infty} a_n x^{n+1}$ integrated term by term and the original power series are equal.

Property 3 (Differentiate Term by Term) Suppose that the interval of convergence of power series $\sum_{n=0}^{\infty} a_n x^n$ is $(-R, R)$, and for any $x \in (-R, R)$, then the sum function $S(x)$ of the power series could differentiate in $(-R, R)$, and for any $x \in (-R, R)$, we have

$$S'(x) = \left(\sum_{n=0}^{\infty} a_n x^n \right)'$$
$$= \sum_{n=0}^{\infty} (a_n x^n)'$$
$$= \sum_{n=1}^{\infty} n a_n x^{n-1}$$
$$= a_1 + 2 a_2 x + 3 a_3 x^2 + \cdots,$$

I 上连续、可导或可积的函数,它们的和在区间 I 上同样连续、可导或可积.这些结论可以推广到无穷项幂函数之和——幂级数上.

性质 1 (和函数的连续性) 设幂级数 $\sum_{n=0}^{\infty} a_n x^n$ 的收敛域 $D \neq \{0\}$,则该幂级数的和函数 $S(x)$ 在 D 上连续.

性质 2 (逐项积分) 设幂级数 $\sum_{n=0}^{\infty} a_n x^n$ 的收敛区间为 $(-R, R)$,则该幂级数的和函数 $S(x)$ 在 $(-R, R)$ 内部可积,且对于任意 $x \in (-R, R)$,有

$$\int_0^x S(x) dx = \int_0^x \left(\sum_{n=0}^{\infty} a_n x^n \right) dx$$
$$= \sum_{n=0}^{\infty} \int_0^x a_n x^n dx$$
$$= \sum_{n=0}^{\infty} \frac{a_n}{n+1} x^{n+1}$$
$$= a_0 x + \frac{1}{2} a_1 x^2 + \frac{1}{3} a_2 x^3 + \cdots,$$

其中逐项积分后得到的幂级数 $\sum_{n=0}^{\infty} \frac{a_n}{n+1} x^{n+1}$ 与原幂级数的收敛半径相同.

性质 3 (逐项求导) 设幂级数 $\sum_{n=0}^{\infty} a_n x^n$ 的收敛区间为 $(-R, R)$,则该幂级数的和函数 $S(x)$ 在 $(-R, R)$ 内可导,且对于任意 $x \in (-R, R)$,有

$$S'(x) = \left(\sum_{n=0}^{\infty} a_n x^n \right)'$$
$$= \sum_{n=0}^{\infty} (a_n x^n)'$$
$$= \sum_{n=1}^{\infty} n a_n x^{n-1}$$
$$= a_1 + 2 a_2 x + 3 a_3 x^2 + \cdots,$$

where the convergence radius of the power series $\sum_{n=1}^{\infty} n a_n x^{n-1}$ differentiated term by term and the original power series is the same.

Using the above properties, we can find the sum functions of some power series, and the sum of some special constant series.

Example 1 Apply the geometric series
$$\frac{1}{1-x} = 1 + x + x^2 + x^3 + \cdots$$
$$(-1 < x < 1)$$
to obtain formulas for two new power series, and find the sum of the series
$$1 - \frac{1}{2} + \frac{1}{3} - \frac{1}{4} + \cdots.$$

Solution Differentiating the geometric series term by term, we get
$$\frac{1}{(1-x)^2} = 1 + 2x + 3x^2 + 4x^3 + \cdots$$
$$(-1 < x < 1).$$

Simultaneously, integrating it term by term, we get
$$\int_0^x \frac{1}{1-t} dt = \int_0^x 1 dt + \int_0^x t dt + \int_0^x t^2 dt + \cdots,$$
that is
$$-\ln(1-x) = x + \frac{x^2}{2} + \frac{x^3}{3} + \cdots$$
$$(-1 < x < 1).$$

If we replace x by $-x$, we obtain
$$\ln(1+x) = x - \frac{x^2}{2} + \frac{x^3}{3} - \frac{x^4}{4} + \cdots$$
$$(-1 < x < 1).$$

When $x = 1$, the series of the right side of the above formula is
$$1 - \frac{1}{2} + \frac{1}{3} - \frac{1}{4} + \cdots = \sum_{n=1}^{\infty} \frac{(-1)^n}{n},$$
which is convergent alternating series. So, by Property 1, we know that
$$\ln 2 = 1 - \frac{1}{2} + \frac{1}{3} - \frac{1}{4} + \cdots.$$

Example 2 Find the power series expression of arctan x.

其中逐项求导后得到的幂级数 $\sum_{n=1}^{\infty} n a_n x^{n-1}$ 与原幂级数的收敛半径相同.

利用上面的性质,可以求出一些幂级数的和函数以及一些特殊常数项级数的和.

例1 利用几何级数
$$\frac{1}{1-x} = 1 + x + x^2 + x^3 + \cdots$$
$$(-1 < x < 1),$$
求两个新的幂级数公式,并求级数
$$1 - \frac{1}{2} + \frac{1}{3} - \frac{1}{4} + \cdots$$
的和.

解 对给出的几何级数逐项求导,我们有
$$\frac{1}{(1-x)^2} = 1 + 2x + 3x^2 + 4x^3 + \cdots$$
$$(-1 < x < 1).$$

再对给出的几何级数逐项积分,我们有
$$\int_0^x \frac{1}{1-t} dt = \int_0^x 1 dt + \int_0^x t dt + \int_0^x t^2 dt + \cdots,$$
即
$$-\ln(1-x) = x + \frac{x^2}{2} + \frac{x^3}{3} + \cdots$$
$$(-1 < x < 1).$$

如果在上式中把 x 替换为 $-x$,我们得到
$$\ln(1+x) = x - \frac{x^2}{2} + \frac{x^3}{3} - \frac{x^4}{4} + \cdots$$
$$(-1 < x < 1).$$

当 $x = 1$ 时,上式右端的级数为
$$1 - \frac{1}{2} + \frac{1}{3} - \frac{1}{4} + \cdots = \sum_{n=1}^{\infty} \frac{(-1)^n}{n},$$
它是收敛的交错级数.于是,由性质1知
$$\ln 2 = 1 - \frac{1}{2} + \frac{1}{3} - \frac{1}{4} + \cdots.$$

例2 求 arctan x 幂级数表示式.

4.5 Operations and Properties of Power Series

Solution Recall that
$$\arctan x = \int_0^x \frac{1}{1+t^2}\,dt.$$

Expand $\frac{1}{1-x}$ to power series, and replaced x by $-t^2$, we get
$$\frac{1}{1+t^2} = 1 - t^2 + t^4 - t^6 + \cdots$$
$$(-1 < t < 1).$$

Thus,
$$\arctan x = \int_0^x (1 - t^2 + t^4 - t^6 + \cdots)\,dt$$
$$(-1 < x < 1),$$

that is
$$\arctan x = x - \frac{x^3}{3} + \frac{x^5}{5} - \frac{x^7}{7} + \cdots$$
$$(-1 < x < 1).$$

Example 3 Find the sum function of the power series
$$1 + x + \frac{x^2}{2!} + \frac{x^3}{3!} + \cdots.$$

Solution We have known that this series convergent for all x. Let the sum function of this series be $S(x)$. Differentiating $S(x)$ term by term, we obtain
$$S'(x) = 1 + x + \frac{x^2}{2!} + \frac{x^3}{3!} + \cdots$$
$$(-\infty < x < +\infty),$$

that is
$$S'(x) = S(x)$$

for all x. Furthermore, $S(0) = 1$. It is easy to get a function $S(x)$ that satisfies these two conditions, that is
$$S(x) = e^x.$$

Thus
$$e^x = 1 + x + \frac{x^2}{2!} + \frac{x^3}{3!} + \cdots$$
$$(-\infty < x < +\infty).$$

Example 4 Find the power series expression of e^{-x^2}.

Solution In the power series expression of e^x obtained from the above example, we could have
$$e^{-x^2} = 1 - x^2 + \frac{x^4}{2!} - \frac{x^6}{3!} + \cdots$$
$$(-\infty < x < +\infty).$$

4.5 幂级数的运算与性质

解 我们先回顾一下：
$$\arctan x = \int_0^x \frac{1}{1+t^2}\,dt.$$

将 $\frac{1}{1-x}$ 展开成幂级数，并将 x 替换为 $-t^2$，我们得到
$$\frac{1}{1+t^2} = 1 - t^2 + t^4 - t^6 + \cdots$$
$$(-1 < t < 1).$$

因此
$$\arctan x = \int_0^x (1 - t^2 + t^4 - t^6 + \cdots)\,dt$$
$$(-1 < x < 1),$$

即
$$\arctan x = x - \frac{x^3}{3} + \frac{x^5}{5} - \frac{x^7}{7} + \cdots$$
$$(-1 < x < 1).$$

例 3 求幂级数
$$1 + x + \frac{x^2}{2!} + \frac{x^3}{3!} + \cdots$$
的和函数.

解 我们已经知道，对所有 x，这个级数都收敛. 设该级数的和函数为 $S(x)$. 通过逐项求导，我们得到
$$S'(x) = 1 + x + \frac{x^2}{2!} + \frac{x^3}{3!} + \cdots$$
$$(-\infty < x < +\infty).$$

也就是说，对于所有的 x，有
$$S'(x) = S(x).$$

进一步，有 $S(0) = 1$. 容易得到满足这两个条件的函数 $S(x)$ 为
$$S(x) = e^x.$$

因此
$$e^x = 1 + x + \frac{x^2}{2!} + \frac{x^3}{3!} + \cdots$$
$$(-\infty < x < +\infty).$$

例 4 求 e^{-x^2} 的幂级数表示式.

解 在上例得到的 e^x 的幂级数表达式中，将 x 替换为 $-x^2$，我们得到
$$e^{-x^2} = 1 - x^2 + \frac{x^4}{2!} - \frac{x^6}{3!} + \cdots$$
$$(-\infty < x < +\infty).$$

Exercises 4

1. Find the sum of the following series:

   (1) $\sum_{n=0}^{\infty} \dfrac{1}{(n+1)(n+2)}$;

   (2) $\sum_{n=1}^{\infty} \dfrac{1}{2n(n+1)}$;

   (3) $\sum_{n=0}^{\infty} \dfrac{3}{10^n}$;

   (4) $\sum_{n=0}^{\infty} \dfrac{2^{n+3}}{3^n}$.

2. Check the following series of convergence:

   (1) $\sum_{n=1}^{\infty} \dfrac{n}{1+n}$;

   (2) $\sum_{n=0}^{\infty} \dfrac{1}{2^{n+3}}$;

   (3) $\sum_{n=1}^{\infty} \left(\dfrac{1}{2^{n-1}} + \dfrac{2^n}{3^{n-1}} \right)$.

3. Determine whether the following series converges or diverges:

   (1) $\sum_{n=0}^{\infty} \dfrac{n}{n^3+1}$;   (2) $\sum_{n=0}^{\infty} \dfrac{1}{(2n+1)^2}$;

   (3) $\sum_{n=0}^{\infty} \dfrac{1}{\sqrt{n+1}}$;   (4) $\sum_{n=1}^{\infty} \dfrac{1}{\sqrt{2n^2-n}}$;

   (5) $\sum_{n=1}^{\infty} \dfrac{1}{n\sqrt[n]{n}}$;   (6) $\sum_{n=1}^{\infty} \dfrac{n!}{2^{n^2}}$;

   (7) $\sum_{n=0}^{\infty} \dfrac{n\cos^2 \frac{n\pi}{3}}{2^n}$;   (8) $\sum_{n=2}^{\infty} \dfrac{1}{\ln^2 n}$.

4. Prove that if positive terms series $\sum_{n=1}^{\infty} a_n$ is convergent, so
$$\lim_{n \to \infty} \dfrac{1}{n}(a_1 + 2a_2 + \cdots + na_n) = 0$$
is founded.

5. Determine whether the following series is absolutely convergent:

(1) $\sum_{n=1}^{\infty} \frac{\sin n}{n^3}$;

(2) $\sum_{n=0}^{\infty} \frac{(-1)^n}{n^2+1}$;

(3) $\sum_{n=1}^{\infty} \frac{(-1)^n + \cos 3^n}{n^2+n}$.

6. Distinguish whether the following series is convergent or not. If it is convergent, is it the absolute convergence or conditional convergence?

(1) $\sum_{n=1}^{\infty} (-1)^{n-1} \frac{1}{\sqrt{n}}$;

(2) $\sum_{n=1}^{\infty} (-1)^{n-1} \frac{1}{8^n n}$;

(3) $\sum_{n=1}^{\infty} (-1)^{n-1} \sin \frac{1}{n^3}$;

(4) $\sum_{n=1}^{\infty} (-1)^{n-1} \ln \frac{n+1}{n}$;

(5) $-\frac{1}{1+a} + \frac{1}{2+a} - \frac{1}{3+a} + \frac{1}{4+a} - \cdots$

(a is not a negative integer);

(6) $\frac{1}{\ln 2} - \frac{1}{\ln 3} + \frac{1}{\ln 4} - \frac{1}{\ln 5} + \cdots$;

(7) $\frac{1}{\pi^2} \sin \frac{\pi}{2} - \frac{1}{\pi^3} \sin \frac{\pi}{3} + \frac{1}{\pi^4} \sin \frac{\pi}{4} - \cdots$;

(8) $\sin \frac{1}{1^2} - \sin \frac{1}{2^2} + \sin \frac{1}{3^2} - \sin \frac{1}{4^2} + \cdots$.

7. Find the interval of convergence of $\sum_{n=1}^{\infty} (-1)^{n-1} \frac{x^n}{n}$.

8. Find the interval of convergence of the power series $\sum_{n=1}^{\infty} (-1)^{n-1} \frac{x^{2n-1}}{2n-1}$.

9. Find the interval of convergence and sum function of the following series:

$$\frac{1}{x} + \frac{1}{3x^3} + \frac{1}{5x^5} + \cdots + \frac{1}{(2n+1)x^{2n+1}} + \cdots.$$

10. Find the radius of convergence and the domain of convergence of the following power series:

(1) $\sum_{n=1}^{\infty} n x^n$;   (2) $\sum_{n=1}^{\infty} \frac{x^n}{n^2 2^n}$;

(3) $\sum_{n=0}^{\infty} x^{n^2}$; (4) $\sum_{n=1}^{\infty} \frac{(x-2)^{2n-1}}{(2n-1)!}$;

(5) $\sum_{n=1}^{\infty} \frac{3^n + (-2)^n}{n}$.

11. Using the power series expansion of the known function, find the power series expansions of the following functions at $x=0$, and the domain of convergence of the function is determined.

(1) e^{x^2}; (2) $\frac{x^{10}}{1-x}$;

(3) $\frac{x}{\sqrt{1-2x}}$; (4) $\frac{e^x}{1-x}$;

(5) $\frac{x}{1+x-2x^2}$.